T0212546

SpringerBriefs in Probability and Mathematical Statistics

More information about this series at http://www.springer.com/series/14353

Leonid Mytnik • Vitali Wachtel

Regularity and Irregularity of Superprocesses with $(1 + \beta)$-stable Branching Mechanism

 Springer

Leonid Mytnik
Industrial Engineering and Management
Technion – Israel Institute of Technology
Technion City, Haifa, Israel

Vitali Wachtel
Mathematical Institute
University of Augsburg
Augsburg, Germany

ISSN 2365-4333   ISSN 2365-4341   (electronic)
SpringerBriefs in Probability and Mathematical Statistics
ISBN 978-3-319-50084-3   ISBN 978-3-319-50085-0   (eBook)
DOI 10.1007/978-3-319-50085-0

Library of Congress Control Number: 2016958567

Mathematics Subject Classification (2010): 60J68, 28A80, 60G52, 60G57

Printed on acid-free paper

This Springer imprint is published by Springer Nature
The registered company is Springer International Publishing AG
The registered company address is: Gewerbestrasse 11, 6330 Cham, Switzerland

Preface

This book is devoted to the regularity and fractal properties of the superprocesses with $(1+\beta)$-stable branching. Regularity properties of functions is one of the fundamental questions in analysis. Trajectories of stochastic processes are a rich source of functions with very interesting local smoothness properties. The most classical example is the Brownian motion: almost every path of the Brownian motion is continuous but nowhere differentiable. In the last several decades a class of functions and measures whose regularity properties may change from point to point attracted attention of many researchers. In this book, we will be dealing with the regularity properties of super-Brownian motion with $(1+\beta)$-stable branching mechanism—a measure-valued stochastic process that arises as a scaling limit of branching Brownian motions whose critical branching mechanism is in the domain of attraction of $(1+\beta)$-stable law with $\beta \in (0,1)$.

It has been well known for a long time that the super-Brownian motion with $(1+\beta)$-stable branching mechanism has densities at any fixed time, provided that the spatial dimension d is small enough $(d < 2/\beta)$. Then, in a paper by Mytnik and Perkins (2003) it was shown that in the case of $\beta < 1$ there is a dichotomy for the corresponding density functions at fixed times: they are either continuous if $d = 1$ or locally unbounded in dimensions $d \in (1, 2/\beta)$. Later, in a series of papers by Fleischmann, Mytnik and Wachtel, the Hölder regularity properties and multifractal spectrum of singularities of the densities were determined in dimension $d = 1$. In particular, it was shown that there are spatial points of different Hölder indexes of continuity. These results are in a striking contrast to those in the case of $\beta = 1$ (binary branching mechanism) where the density of super-Brownian motion is "monofractal": it has Hölder index of spatial regularity equal to $1/2$ at every point where it is positive.

This book is a comprehensive and self-contained introduction to the fascinating area of the regularity properties of the densities of superprocesses. It describes the state of the art in this area and gives an overview of the above mentioned results together with main proofs. It also contains some new results that nicely complement the existing ones.

We would like to express our thanks to our friend Klaus Fleischmann who has played a crucial role in the projects on regularity properties of super-Brownian motions: many of our joint results are described in this book.

Haifa, Israel   Leonid Mytnik
Augsburg, Germany   Vitali Wachtel
December 2016

Contents

Chapter 1
Introduction, main results, and discussion

1.1 Model and motivation

This book is devoted to regularity and fractal properties of superprocesses with $(1 + \beta)$-branching. Regularity properties of functions is the most classical question in analysis. Typically one is interested in such properties as continuity/discontinuity and differentiability. Starting from Weierstrass, who constructed an example of a continuous but nowhere differentiable function, people got more and more interested in such "strange" properties of functions. Trajectories of stochastic processes give a rich source of such functions. The most classical example is the Brownian motion: almost every path of the Brownian motion is continuous but nowhere differentiable.

In order to measure the regularity of a function f at point x_0, we need to introduce the so-called Hölder classes $C^\eta(x_0)$. One says that $f \in C^\eta(x_0)$, $\eta > 0$ if there exists a polynomial P of degree $[\eta]$ such that

$$|f(x) - P(x - x_0)| = O(|x - x_0|^\eta).$$

For $\eta \in (0, 1)$ the above definition coincides with the definition of Hölder continuity with index η at x_0. With the definition of $f \in C^\eta(x_0)$ at hand, let us define the *pointwise Hölder exponent* of f at x_0:

$$H_f(x_0) := \sup\{\eta > 0 : f \in C^\eta(x_0)\}, \tag{1.1}$$

and we set it to 0 if $f \notin C^\eta(x_0)$ for all $\eta > 0$. To simplify the exposition we will sometimes call $H_f(x_0)$ the Hölder exponent of f at x_0.

It is well known that the Weierstrass functions have the same Hölder exponent at all points. The same is true for the Brownian motion: the pointwise Hölder exponent at all times is equal to $1/2$ with probability one. However, there exist functions with Hölder exponent changing from point to point. In such a case one speaks of a *multifractal* function (see Frisch and Parisi [20]). For some classical examples of

© The Author(s) 2016
L. Mytnik, V. Wachtel, *Regularity and Irregularity of Superprocesses with (1 + β)-stable Branching Mechanism*, SpringerBriefs in Probability and Mathematical Statistics, DOI 10.1007/978-3-319-50085-0_1

deterministic multifractal functions we refer to Jaffard [27]. In studying multifractal functions people are interested in the "size" of the set of points with given Hölder exponent. To measure these sizes for different Hölder exponents of a function f one introduces the following function:

$$D(\eta) = \dim\{x_0 : H_f(x_0) = \eta\}, \tag{1.2}$$

where $\dim(A)$ denotes the Hausdorff dimension of the set A. The mapping $\eta \mapsto D(\eta)$ reveals the so-called *multifractal spectrum* related to pointwise Hölder exponents of f. A standard example of a multifractal random function is given by a Lévy process with infinite Lévy measure. Its multifractal spectrum was determined by Jaffard in [28]. In general, the multifractal analysis of random functions and measures has attracted attention for many years. It has been studied, for example, in Dembo et al. [12], Durand [13], Hu and Taylor [24], Klenke and Mörters [29], Durand and Jaffard [14]. These examples are related to Lévy-type processes. One could also mention *multifractional* processes that often also give rise to multifractal functions, see, e.g., Balança [1, 2], Le Guével and Lévy-Véhel [34], Riedi [42] among many other references. Jump processes whose jump measure changes according to the position of the process (see, e.g., Barral et al. [8], Yang [46]) and processes in multifractal time (see, e.g., Barral and Seuret [7]) also have interesting multifractal properties.

As for the multifractal structure of random measures, it has also been extensively studied in the literature; for example, it has been investigated for random Gibbs measures and random cascades: here we could mention Holley and Waymire [23], Molchan [35], Barral and Seuret [5, 6], Barral et al. [9] among many works. Another set of examples of random measures having rich multifractal structure arises from random measures defined on branching random trees and it has been studied by Mörters and Shieh [36], Berestycki, Berestycki and Schweinsberg [10], and more recently by Balança [3]. The analysis of multifractal spectrum has also been done in some variations for measure-valued branching processes, see, e.g., Le Gall and Perkins [33], Perkins and Taylor [40], and more recent works: Mytnik and Wachtel [38], Balança and Mytnik [4]. One of the aims of this review is to describe results on multifractal spectrum and other regularity properties of measure-valued branching processes, and especially their density functions, and also show the methods of proofs. We shall do it in the particular case of $(1 + \beta)$-stable super-Brownian motion, whose densities in low dimensions turn out to have a very non-trivial regularity structure. For standard texts on superprocesses we refer to Dawson [11], Etheridge [16], and Perkins [39].

Before we start with the precise definition of these processes, we need to introduce the following notation. $\mathcal{M}$ is the space of all Radon measures on $\mathbb{R}^d$ and $\mathcal{M}_f$ is the space of finite measures on $\mathbb{R}^d$ with weak topology ($\Rightarrow$ denotes weak convergence). In general if F is a set of functions, write F_+ or F^+ for non-negative functions in F. For any metric space E, let $\mathscr{C}_E$ (respectively, $\mathscr{D}_E$) denote the space of continuous (respectively, càdlàg) E-valued paths with compact-open (respectively,

Skorokhod) topology. The integral of a function ϕ with respect to a measure μ is written as $\langle \mu, \phi \rangle$ or $\langle \phi, \mu \rangle$ or $\mu(\phi)$. We use c (or C) to denote a positive, finite constant whose value may vary from place to place. A constant of the form $c(a,b,\dots)$ means that this constant depends on parameters $a,b,\dots$. Moreover, $c_{(\#)}$ will denote a constant appearing in formula line (or array) (#).

Let $(\Omega, \mathscr{F}_t, \mathscr{F}, \mathbf{P})$ be the probability space with filtration, which is sufficiently large to contain all the processes defined below. Let $\mathscr{C}(E)$ denote the space of continuous functions on E and let $\mathscr{C}_b(E)$ be the space of bounded functions in $\mathscr{C}(E)$. Let $\mathscr{C}_b^n = \mathscr{C}_b^n(\mathbb{R}^d)$ denote the subspace of functions in $\mathscr{C}_b = \mathscr{C}_b(\mathbb{R}^d)$ whose partial derivatives of order n or less are also in $\mathscr{C}_b$. A càdlàg adapted measure-valued process X is called a super-Brownian motion with $(1+\beta)$-stable branching if X satisfies the following martingale problem. For every $\varphi \in \mathscr{C}_b^2$ and every $f \in \mathscr{C}^2(\mathbb{R})$,

$$f(\langle X_t, \varphi \rangle) - f(\langle X_0, \varphi \rangle) - \int_0^t f'(\langle X_s, \varphi \rangle)\langle X_s, \tfrac{1}{2}\Delta\varphi \rangle ds \tag{1.3}$$

$$- \int_0^t \left(\int_{\mathbb{R}^d} \int_{(0,\infty)} \left(f(\langle X_s, \varphi \rangle + r\varphi(x)) - f(\langle X_s, \varphi \rangle) - f'(\langle X_s, \varphi \rangle)r\varphi(x) \right) n(dr)X_s(dx) \right) ds$$

is an $\mathscr{F}_t$-martingale, where

$$n(dr) = \frac{\beta(\beta+1)}{\Gamma(1-\beta)} r^{-2-\beta} dr. \tag{1.4}$$

There is also an analytic description of this process: For every positive $\varphi \in \mathscr{C}_b^2$ one has

$$\mathbf{E}e^{-\langle X_t, \varphi \rangle} = e^{-\langle X_0, u_t \rangle}, \tag{1.5}$$

where u is the solution to the log-Laplace equation

$$\frac{d}{dt}u = \frac{1}{2}\Delta u - u^{1+\beta}, \tag{1.6}$$

with the initial condition φ.

If $\beta = 1$, X has continuous sample $\mathscr{M}_f$-valued paths, while for $0 < \beta < 1$, X is a.s. discontinuous and has jumps all of the form $\Delta X_t = \delta_{x(t)}m(t)$ and the set of jump times is dense in $[0, \zeta)$, where $\zeta = \inf\{t : \langle X_t, 1 \rangle = 0\}$ is the lifetime of X (see, for example, Section 6.2.2 of [11]). For $t > 0$ fixed, X_t is absolutely continuous a.s. if and only if $d < 2/\beta$ (see [17] and Theorem 8.3.1 of [11]). If $\beta = 1$, and $d = 1$, then much more can be said — X_t is absolutely continuous for all $t > 0$ a.s. and has a density $X(t,x)$ which is jointly continuous on $(0,\infty) \times \mathbb{R}$ (see [30, 41]). In view of the jumps of X (described above) if $0 < \beta < 1$, we see that X_t cannot have a density for a dense set of times a.s. and the regularity properties of the densities are very intriguing. In this work we consider the "stable branching" case of $0 < \beta < 1$ and consider the question:

What are the regularity properties of the density of X at fixed times t?

The analytic methods used in [17] to prove the existence of a density at a fixed time do not shed any light on its regularity properties. However recently new techniques have been developed that allowed to treat these questions. To the best of our knowledge, the regularity properties of the densities for super-Brownian motion with β-stable branching were first studied in Mytnik and Perkins [37]. It was shown there that there is a continuous version of the density if and only if $d = 1$. Moreover, when $d > 1$ the density is very badly behaved. Note that in the case of $\beta = 1$, the density of super-Brownian motion $X_t(dx)$ exists only in dimension $d = 1$ and the density has a version which is Hölder continuous with any exponent smaller than $1/2$ (see Konno and Shiga [30]).

Now consider the case $\beta < 1$. In a series of papers of Fleischmann, Mytnik, and Wachtel [18, 19] and Mytnik and Wachtel [38] the properties of the density were studied for a superprocess with β-stable branching with an α-stable motion, the so-called (α, d, β)-superprocess. To be more precise, (α, d, β)-superprocess is the measure-valued process whose Laplace transform is described by (1.5), whereas u solves (1.6) with $\frac{1}{2}\Delta$ replaced by the fractional Laplacian $-(-\Delta)^{\alpha/2}$.

As we have mentioned above, first regularity properties of the densities of superprocesses with β-stable branching were derived by Mytnik and Perkins [37] for $(2, d, \beta)$-superprocesses. In [18], these results were extended to the case of (α, d, β)-superprocesses with arbitrary $\alpha \in (0, 2]$. In particular, it was shown that there is a *dichotomy* for the density function of the measure (in what follows, we just say the "density of the measure"): There is a continuous version of the density of $X_t(dx)$ if $d = 1$ and $\alpha > 1 + \beta$, but otherwise the density is unbounded on open sets of positive $X_t(dx)$-measure. Moreover, in the case of continuity ($d = 1$ and $\alpha > 1 + \beta$), Hölder regularity properties of the density had been studied in [18, 19, 38]. It turned out that on any set of positive $X_t(dx)$ measure, there are points with different pointwise Hölder exponents. In [38] the Hausdorff dimensions of sets containing the points with certain Hölder exponents were computed: this reveals the multifractal spectrum related to pointwise Hölder exponents.

The main purpose of this book is to give concise exposition of the results on the regularity properties of densities of superprocesses with stable branching. On top of it we will also prove some new results that give a more complete picture of regularity properties.

1.2 Results on regularity properties of the densities of super-Brownian motion with stable branching

As we have mentioned above we are interested in the regularity properties of the (α, d, β)-superprocess with $\beta \in (0, 1)$. In this book we will consider the particular case of

$$\alpha = 2, \tag{1.7}$$

that is, the case of super-Brownian motion. We do it in order to simplify the exposition; however, the proofs go through also in the case of α-stable motion process. The enthusiastic reader who is interested in this general case is invited to go through the series of papers [18, 19, 38].

So, from now on we assume (1.7) and

$$\beta < 1.$$

Let us introduce the following notation: for any function $f : \mathbb{R}^d \mapsto \mathbb{R}$ and any open set $B \subset \mathbb{R}^d$ denote

$$\|f\|_B := \operatorname*{ess\,sup}_{x \in B} |f(x)|.$$

The first result deals with the *dichotomy* of the density of super-Brownian motion, see [37] and [18]. Recall that, by [17], the density for fixed times $t > 0$, exists if and only if $d < 2/\beta$.

Theorem 1.1 (Dichotomy for densities). *Let $d < 2/\beta$. Fix $t > 0$ and $X_0 = \mu \in \mathcal{M}_{\mathrm{f}}$.*

(a) *If $d = 1$ then with probability one, there is a continuous version $\tilde{X}_t$ of the density function of the measure $X_t(\mathrm{d}x)$.*

(b) *If $d > 1$, then with probability one, for all open $U \subseteq \mathbb{R}^d$,*

$$\|X_t\|_U = \infty \ \text{whenever} \ X_t(U) > 0.$$

Convention. For the rest of the manuscript, if $d = 1$ and t is fixed, then with a slight abuse of notation, $X_t(x)$, $x \in \mathbb{R}$ will denote the continuous version of the density of the measure $X_t(\mathrm{d}x)$. Such a version exists due to the previous theorem.

For the later results in this subsection we assume that $d = 1$. In the next theorem the first regularity properties of the density $X_t(x)$, $x \in \mathbb{R}$, are revealed (see [18]).

Theorem 1.2 (Local Hölder continuity). *Let $d = 1$. Fix $t > 0$ and $X_0 = \mu \in \mathcal{M}_{\mathrm{f}}$.*

(a) *For each $\eta < \eta_{\mathrm{c}} := \frac{2}{1+\beta} - 1$, $X_t(\cdot)$ is locally Hölder continuous of index η :*

$$\sup_{x_1, x_2 \in K, x_1 \neq x_2} \frac{|X_t(x_1) - X_t(x_2)|}{|x_1 - x_2|^\eta} < \infty, \quad \text{for any compact } K \subset \mathbb{R}.$$

(b) *For every $\eta \geq \eta_{\mathrm{c}}$ with probability one, for any open $U \subseteq \mathbb{R}$,*

$$\sup_{x_1, x_2 \in U, x_1 \neq x_2} \frac{|X_t(x_1) - X_t(x_2)|}{|x_1 - x_2|^\eta} = \infty \quad \text{whenever } X_t(U) > 0.$$

One of the consequences of the above theorem is that the so-called *optimal index* for *local* Hölder continuity of X_t equals to η_{c} (see, e.g., Section 2.2 in [44] for the discussion about local Holder index of continuity).

The main part of this book is devoted to studying pointwise regularity properties of X_t which differ drastically from their local regularity properties. In particular, we are interested in pointwise Hölder exponent at fixed points and in multifractal spectrum.

In what follows, $H_X(x)$ will denote the pointwise Hölder exponent of the density function X_t at $x \in \mathbb{R}$. The next result describes $H_X(x)$ at *fixed* points $x \in \mathbb{R}$.

Theorem 1.3 (Pointwise Hölder exponent at fixed points). *Let* $d = 1$. *Fix* $t > 0$ *and* $X_0 = \mu \in \mathcal{M}_f$. *Define* $\bar{\eta}_c := \frac{3}{1+\beta} - 1$. *If* $\bar{\eta}_c \neq 1$, *then for every fixed* $x \in \mathbb{R}$,

$$H_X(x) = \bar{\eta}_c, \quad \mathbf{P} - a.s. \text{ on } \{X_t(x) > 0\}.$$

Remark 1.4. The above result was proved in [19] for the case of $\beta > 1/2$. This is the case for which $H_X(x) < 1$. In the case of $\beta \leq 1/2$, it was shown in [19], that for any fixed point $x \in \mathbb{R}$

$$H_X(x) \geq 1 \quad \mathbf{P} - a.s. \text{ on } \{X_t(x) > 0\}.$$

Thus, Theorem 1.3 strengthens the result from [19] by determining the pointwise Hölder exponent at fixed points for any $\beta \in (0,1) \setminus \{\frac{1}{2}\}$. The case $\beta = 1/2$ corresponds to $\bar{\eta}_c = 1$ (integer-valued). This creates serious technical difficulties, which are related to the polynomial correction term in the definition of the pointwise Hölder exponent.

The above results immediately imply that almost every realization of X_t has points with different pointwise exponents of continuity. For example, it follows from Theorem 1.2 that one can find (random) points x where $H_X(x) = \eta_c$. Moreover, it follows from Theorem 1.3 that there are also points x where $H_X(x) = \bar{\eta}_c > \eta_c$. This indicates that we are dealing with random multifractal function $x \mapsto X_t(x)$. To study its multifractal spectrum, for any open $U \subset \mathbb{R}$ and any $\eta \in (\eta_c, \bar{\eta}_c]$ define a random set

$$\mathcal{E}_{U,X,\eta} := \{x \in U : H_X(x) = \eta\}$$

and let $D_U(\eta)$ denote its Hausdorff dimension (similarly to (1.2)).

The function $\eta \mapsto D_U(\eta)$ reveals the multifractal spectrum related to pointwise Hölder exponents of $X_t(\cdot)$. This spectrum is determined in the next theorem (see [38]) which also claims its independence on U.

Theorem 1.5 (Multifractal spectrum). *Fix* $t > 0$, *and* $X_0 = \mu \in \mathcal{M}_f$. *Let* $d = 1$. *Then, for any* $\eta \in [\eta_c, \bar{\eta}_c] \setminus \{1\}$ *and any open set* U, *with probability one*,

$$D_U(\eta) = (\beta + 1)(\eta - \eta_c) \tag{1.8}$$

whenever $X_t(U) > 0$.

It should be emphasized that the result in Theorem 1.5 is not uniform in η. More precisely, an event of zero probability, on which (1.8) can fail, is not necessarily the same for different values of the exponent η. The question, whether there exists a zero set M such that (1.8) holds for all η and all $\omega \in M^c$, remains open. However, the upper bound on D_U can be made uniform in η:

Proposition 1.6. *Fix* $t > 0$, *and* $X_0 = \mu \in \mathcal{M}_f$. *Let* $d = 1$. *Then, for any open set* U, *with probability one,*

$$D_U(\eta) \leq (\beta + 1)(\eta - \eta_c), \quad \text{for all } \eta \in [\eta_c, \bar{\eta}_c],$$

whenever $X_t(U) > 0$.

Note that the uniformity of multifractal spectrum in η has been obtained by Balança [3] for closely related model — this has been done for level sets of stable random trees. In Balança and Mytnik [4] such uniformity was obtained for the spectrum of singularities (for measures) of $(2,d,\beta)$-superprocesses.

Remark 1.7. The proof of Theorem 1.5 fails in the case $\eta = 1$, and it is a bit disappointing. Formally, it happens for some technical reasons, but one has also to note that this point is critical: it is the borderline between differentiable and non-differentiable functions, see also Remark 1.4. As one can infer from Proposition 1.6, only the lower bound on $D_U(\eta)$ is problematic for $\eta = 1$. However, we still believe that the function $D_U(\cdot)$ can be continuously extended to $\eta = 1$, i.e., $D_U(1) = (\beta + 1)(1 - \eta_c)$ almost surely on $\{X_t(U) > 0\}$.

Remark 1.8. The condition $\beta < 1$ for the above results excludes the case of the quadratic super-Brownian motion, i.e., $\beta = 1$. But it is a known "folklore" result that the super-Brownian motion $X_t(\cdot)$ is almost surely monofractal on any open set of strictly positive density. That is, **P**-a.s., for any x with $X_t(x) > 0$ we have $H_X(x) = 1/2$. For the fact that $H_X(x) \geq 1/2$, for any x, see Konno and Shiga [30] and Walsh [45]. To get that $H_X(x) \leq 1/2$ on the event $\{X_t(x) > 0\}$ one can show that

$$\limsup_{\delta \to 0} \frac{|X_t(x+\delta) - X_t(x)|}{\delta^\eta} = \infty \quad \text{for all } x \text{ such that } X_t(x) > 0, \ \mathbf{P}-\text{a.s.},$$

for every $\eta > 1/2$. This result follows from the fact that for $\beta = 1$ the noise driving the corresponding stochastic equation for X_t is Gaussian (see (0.4) in [30]) in contrast to the case of $\beta < 1$ considered here, where we have driving discontinuous noise with Lévy-type intensity of jumps.

Organization of the article. Beyond the description of the regularity properties of the densities of $(2,d,\beta)$-superprocesses, which is given above, one of the main goals of the article is to provide the approach for proving these properties. We will also show how to use this approach to verify some of the results mentioned above. In particular we will give main elements of the proofs of Theorems 1.1, 1.3, 1.5 and Proposition 1.6. As for the missing details and the proof of Theorem 1.2 we refer the reader to corresponding papers.

Now we will say a few of words about the organization of the material in the following chapters. In Chapter 2, we give the representation of $(2, d, \beta)$-superprocess as a solution to certain martingale problem and describe the approach for studying the regularity of the superprocess. Chapter 3 collects certain properties of $(2, d, \beta)$-superprocesses which later are used for the proofs. Chapter 4 is very important: it gives precise estimates on the sizes of jumps of $(2, d, \beta)$-superprocesses, which in turn are crucial for deriving the regularity properties. Chapters 5, 6, 7 are devoted to the partial proofs of Theorems 1.1, 1.3, 1.5 and Proposition 1.6. Since many of the proofs are technical, at the beginning of several chapters and sections we give heuristic explanations of our results, which, as we hope, will provide the reader with some intuition about the results and their proofs. In the end we have two appendices, where we collect some technical results on transition kernel of the Brownian motion and on spectrally positive stable processes. We use them quite often in the proofs of our main results.

Chapter 2
Stochastic representation for X and description of the approach for determining regularity

Let X be a $(2, d, \beta)$-superprocess, that is, it satisfies the martingale problem (1.3).

The following lemma contains a semimartingale decomposition of X which includes stochastic integrals with respect to discontinuous martingale measures.

Lemma 2.1. *Fix $X_0 = \mu \in \mathcal{M}_f$.*

(a) (Discontinuities)   Define the random measure

$$\mathcal{N} := \sum_{s \in J} \delta_{(s, \Delta X_s)}, \tag{2.1}$$

where J denotes the set of all jump times of X. Then there exists a random counting measure $N\big(\mathrm{d}(s, x, r)\big)$ on $\mathbb{R}_+ \times \mathbb{R}^d \times \mathbb{R}_+$ such that

$$\int_{\mathbb{R}_+} \int_{\mathcal{M}_f} G(s, \mu) \, \mathcal{N}(ds, d\mu) = \int_0^\infty \int_{\mathbb{R}_+} \int_{\mathbb{R}^d} G(s, r\delta_x) N\big(\mathrm{d}(s, x, r)\big), \tag{2.2}$$

for any bounded continuous G on $\mathbb{R}_+ \times \mathcal{M}_f$. That is, all discontinuities of the process X are jumps upwards of the form $r\delta_x$.

(b) (Jump intensities)   The compensator $\widehat{N}$ of N is given by

$$\widehat{N}\big(\mathrm{d}(s, x, r)\big) = ds \, X_s(\mathrm{d}x) \, n(\mathrm{d}r), \tag{2.3}$$

that is, $\widetilde{N} := N - \widehat{N}$ is a martingale measure on $\mathbb{R}_+ \times \mathbb{R}^d \times \mathbb{R}_+$.

(c) (Martingale decomposition)   For all $\varphi \in \mathscr{C}_b^{2,+}$ and $t \geq 0$,

$$\langle X_t, \varphi \rangle = \langle \mu, \varphi \rangle + \int_0^t ds \left\langle X_s, \frac{1}{2} \Delta \varphi \right\rangle + M_t(\varphi) \tag{2.4}$$

© The Author(s) 2016
L. Mytnik, V. Wachtel, *Regularity and Irregularity of Superprocesses with (1 + β)-stable Branching Mechanism*, SpringerBriefs in Probability and Mathematical Statistics, DOI 10.1007/978-3-319-50085-0_2

with discontinuous martingale

$$t \mapsto M_t(\varphi) := \int_0^t \int_{\mathcal{M}_{\mathfrak{f}}} \langle \mu, \varphi \rangle (\mathcal{N} - \widehat{\mathcal{N}}) (d(ds, d\mu))$$

$$= \int_{(0,t] \times \mathbb{R}^d \times \mathbb{R}_+} r \varphi(x) \widetilde{N} (d(s, x, r)) \qquad (2.5)$$

The martingale decomposition of X in the above lemma is basically proven in Dawson [11, Section 6.1]. However, for the sake of completeness we will reprove it here. Some ideas are taken also from [31].

Proof. Since X satisfies the martingale problem (1.3) one can easily get (by formally taking $f(x) = x$) that

$$\langle X_t, \varphi \rangle = \langle \mu, \varphi \rangle + \int_0^t ds \left\langle X_s, \frac{1}{2} \Delta \varphi \right\rangle + \widetilde{M}_t(\varphi)$$

where $\widetilde{M}_t(\varphi)$ is a local martingale. Moreover, by taking again $f(\langle X_t, \varphi \rangle)$ for $f \in \mathscr{C}_b^2(\mathbb{R})$, applying the Itô formula and comparing the terms with (1.3) one can easily see that for each $\varphi \in \mathscr{C}_b^{2,+}(\mathbb{R}^d)$, $\widetilde{M}_t(\varphi)$ is a purely discontinuous martingale, with the compensator measure $\widehat{\mathcal{N}_\varphi}$ given by

$$\int_0^t \int_{\mathbb{R}_+} f(s, v) \widehat{\mathcal{N}_\varphi}(ds, dv) = \int_0^t \int_{\mathbb{R}_+} \int_{\mathbb{R}^d} f(s, r\varphi(x)) X_s(dx) n(dr) \, ds, \, t \geq 0 \quad (2.6)$$

for any bounded continuous $f : \mathbb{R}_+ \times \mathbb{R}_+ \mapsto \mathbb{R}$. This means that if s is a jump time for $\langle X, \varphi \rangle$, then

$$\Delta \langle X_s, \varphi \rangle = \Delta \widetilde{M}_s(\varphi) = r\varphi(x) \qquad (2.7)$$

for some r and $x \in \mathbb{R}^d$ ("distributed according to $\rho X_{s-}(dx)n(dr)$). Since this holds for any test function φ, by putting all together we infer that if $s > 0$ is the jump time for the measure-valued process X, then $\Delta X_s = r\delta_x$ for some r and $x \in \mathbb{R}^d$.

Let $\mathcal{N}$ be as in (2.1). Then by above description of jumps of X, it is clear that there exists a random counting measure N such that

$$N := \sum_{(s,x,r):s \in J, \Delta X_s = r\delta_x} \delta_{(s,x,r)}, \qquad (2.8)$$

and **(a)** follows.

In order to obtain (2.3), we first get $\widehat{\mathcal{N}}$ — the compensator of $\mathcal{N}$. It is defined as follows. For any non-negative predictable function F on $\mathbb{R}_+ \times \Omega \times \mathcal{M}_{\mathfrak{f}}$, $\widehat{\mathcal{N}}$ satisfies the following quality:

$$\mathbf{E}_\mu \left[\int_{\mathbb{R}_+} \int_{\mathcal{M}_{\mathfrak{f}}} F(s, \omega, \mu) \mathcal{N}(ds, d\mu) \right] = \mathbf{E}_\mu \left[\int_{\mathbb{R}_+} \int_{\mathcal{M}_{\mathfrak{f}}} F(s, \omega, \mu) \widehat{\mathcal{N}}(ds, d\mu) \right]. \quad (2.9)$$

We will show that, in fact, $\widehat{\mathcal{N}}$ is defined by the equality

$$\int_{\mathbb{R}_+}\int_{\mathcal{M}_f} G(s,\mu)\,\widehat{\mathcal{N}}(ds,d\mu) = \int_0^\infty ds \int_{\mathbb{R}_+} n(dr)\int_{\mathbb{R}^d} X_s(dx)\,G(s,r\delta_x), \quad (2.10)$$

which holds for any bounded continuous function G on $\mathbb{R}_+ \times \mathcal{M}_f$. To show (2.10), for any bounded non-negative continuous function φ define random measure

$$\mathcal{N}_\varphi := \sum_{s\in J}\delta_{(s,\Delta\langle X_s,\varphi\rangle)},$$

By (2.7) we have that the compensator measure of $\mathcal{N}_\varphi$ is $\widehat{\mathcal{N}}_\varphi$. Clearly for any bounded continuous $f:\mathbb{R}_+ \times \mathbb{R}_+ \mapsto \mathbb{R}$, and $\varphi \in \mathscr{C}_b^{2,+}$

$$\int_0^t \int_{\mathcal{M}_f} f(s,\langle \mu,\varphi\rangle)\,\mathcal{N}(ds,d\mu) = \int_0^t \int_{\mathbb{R}_+} f(s,v)\,\mathcal{N}_\varphi(ds,dv), \quad \forall t \geq 0,$$

This immediately implies that the corresponding compensator measures also satisfy

$$\int_0^t \int_{\mathcal{M}_f} f(s,\langle \mu,\varphi\rangle)\,\widehat{\mathcal{N}}(ds,d\mu)$$

$$= \int_0^t \int_{\mathbb{R}_+} f(s,v)\,\widehat{\mathcal{N}}_\varphi(ds,dv)$$

$$= \int_0^t \int_{\mathbb{R}_+}\int_{\mathbb{R}^d} f(s,r\varphi(x))\,X_s(dx)n(dr)\,ds, \quad (2.11)$$

where the second equality follows by (2.6). Since collection of functions in the form $f(s,\langle \mu,\varphi\rangle)$ is dense in the space of bounded continuous functions on $\mathbb{R}_+ \times \mathcal{M}_f$, (2.10) follows. Now (2.3) follows by **(a)** and the definition of N. This finishes the proof of **(b)**.

To show (2.5) it is enough to derive only the first inequality since the second one is immediate by the definition of N. Let us first identify the class of functions for which the stochastic integral with respect to $(\mathcal{N} - \widehat{\mathcal{N}})(ds,d\mu)$ is well defined. Let F be a measurable function on $\mathbb{R}_+ \times \mathcal{M}_f$ such that, for every $t \geq 0$,

$$\mathbf{E}_\mu\left[\left(\sum_{s\in J\cap[0,t]} F(s,\Delta X_s)^2\right)^{1/2}\right] < \infty. \quad (2.12)$$

Following [26] (Section II.1d), we can then define the stochastic integral of F with respect to the compensated measure $\mathcal{N} - \widehat{\mathcal{N}}$,

$$\int_0^t F(s,\mu)\,(\mathcal{N} - \widehat{\mathcal{N}})(ds,d\mu),$$

as the unique purely discontinuous martingale (vanishing at time 0) whose jumps are indistinguishable of the process $1_J(s)\,F(s,\Delta X_s)$.

We shall be interested in the special case where

$$F(s,\mu) = F_\phi(s,\mu) \equiv \int \phi(s,x)\mu(dx) \qquad (2.13)$$

for some measurable function ϕ on $\mathbb{R}_+ \times \mathbb{R}^d$ (some convention is needed when $\int |\phi(s,x)|\mu(dx) = \infty$, but this will be irrelevant in what follows). If ϕ is bounded, then it is easy to see that condition (2.12) holds. Indeed, we can bound separately

$$\mathbf{E}_\mu\left[\left(\sum_{s\le t}\langle\Delta X_s,1\rangle^2 1_{\{\langle\Delta X_s,1\rangle\le 1\}}\right)^{1/2}\right] \le \mathbf{E}_\mu\left[\sum_{s\le t}\langle\Delta X_s,1\rangle^2 1_{\{\langle\Delta X_s,1\rangle\le 1\}}\right]^{1/2}$$

$$= \left(\int_{(0,1]} r^2 n(dr)\mathbf{E}_\mu\left[\int_0^t\langle X_s,1\rangle ds\right]\right)^{1/2} < \infty,$$

and, using the simple inequality $a_1^2+\cdots+a_n^2 \le (a_1+\cdots+a_n)^2$ for any non-negative reals $a_1,\ldots,a_n$,

$$\mathbf{E}_\mu\left[\left(\sum_{s\le t}\langle\Delta X_s,1\rangle^2 1_{\{\langle\Delta X_s,1\rangle>1\}}\right)^{1/2}\right] \le \mathbf{E}_\mu\left[\sum_{s\le t}\langle\Delta X_s,1\rangle 1_{\{\langle\Delta X_s,1\rangle>1\}}\right]$$

$$= \int_{(1,\infty)} rn(dr)\mathbf{E}_\mu\left[\int_0^t\langle X_s,1\rangle ds\right] < \infty.$$

In both cases, we have used (2.9) and the fact that $\mathbf{E}_\mu[\langle X_t,1\rangle] \le \langle\mu,1\rangle$.

To simplify notation, we write

$$M_t(\phi) = \int_0^t\int_{\mathbb{R}^d}\phi(s,x)M(ds,dx) \equiv \int_0^t F_\phi(s,\mu)(\mathcal{N}-\widehat{\mathcal{N}})(ds,d\mu),$$

whenever (2.12) holds for $F = F_\phi$, where F_ϕ is defined in (2.13). This is consistent with the notation of the introduction. Indeed, if $\phi(s,x) = \varphi(x)$ where $\varphi \in \mathscr{C}_b^2(\mathbb{R})$, then by the very definition, $M_t(\phi)$ is a purely discontinuous martingale with the same jumps as the process $\langle X_t,\varphi\rangle$. Since the same holds for the process

$$\widetilde{M}_t(\varphi) = \langle X_t,\varphi\rangle - \langle X_0,\varphi\rangle - \int_0^t\langle X_s,\tfrac{1}{2}\Delta\varphi\rangle ds$$

(see Theorem 7 in [15]) we get that $M_t(\phi) = \widetilde{M}_t(\varphi)$.   □

Let $\{p_t(x), t\ge 0, x\in\mathbb{R}^d\}$ denote the continuous transition kernel related to the Laplacian Δ in $\mathbb{R}^d$, and $(S_t, t\ge 0)$ the related semigroup, that is,

$$S_t f(x) = \int_{\mathbb{R}^d} p_t(x-y)f(y)dy \text{ for any bounded function } f$$

and

$$S_t\nu(x) = \int_{\mathbb{R}^d} p_t(x-y)\nu(dy) \text{ for any finite measure } \nu.$$

Fix $X_0 = \mu \in \mathscr{M}_f \setminus \{0\}$. Recall that if $d = 1$, then $X_t(dx)$ is a.s. absolutely continuous for every fixed $t > 0$ (see [17]). In what follows till the end of the chapter we will consider the case of $d = 1$. From the Green function representation related to (2.4) (see, e.g., [18, (1.9)]) we obtain the following representation of a version of the density function of $X_t(dx)$ in $d = 1$ (see, e.g., [18, (1.12)]):

$$X_t(x) = \mu * p_t(x) + \int_{(0,t] \times \mathbb{R}} M\big(\mathrm{d}(u,y)\big) p_{t-u}(y - x)$$
$$=: \mu * p_t(x) + Z_t(x), \quad x \in \mathbb{R}, \tag{2.14}$$

where

$$Z_s(x) = \int_{(0,s] \times \mathbb{R}} M\big(\mathrm{d}(u,y)\big) p_{t-u}(y - x), \ 0 \le s \le t. \tag{2.15}$$

Note that although $\{Z_s\}_{s \le t}$ depends on t, it does not appear in the notation since t is fixed throughout the book. $M\big(\mathrm{d}(s,y)\big)$ in (2.14) is the martingale measure related to (2.5). Note that by Lemma 1.7 of [18] the class of "legitimate" integrands with respect to the martingale measure $M\big(\mathrm{d}(s,y)\big)$ includes the set of functions ψ such that for some $p \in (1+\beta, 2)$,

$$\int_0^T ds \int_{\mathbb{R}} dx S_s \mu(x) |\psi(s,x)|^p < \infty, \quad \forall T > 0. \tag{2.16}$$

We let $\mathscr{L}_{\mathrm{loc}}^p$ denote the space of equivalence classes of measurable functions satisfying (2.16). For $\beta < 1$, it is easy to check that, for any $t > 0, z \in \mathbb{R}$,

$$(s,x) \mapsto p_{t-s}(z - x) 1_{s<t}$$

is in $\mathscr{L}_{\mathrm{loc}}^p$ for any $p \in (1+\beta, 2)$, and hence the stochastic integral in the representation (2.14) is well defined.

$\mu * p_t(x)$ is obviously twice differentiable. Thus, the regularity properties of $X_t(\cdot)$ including its multifractal structure coincide with that of Z. Recalling the definitions of Z and $M(ds, dy)$, we see that there is a "competition" between branching and motion: jumps of the martingale measure M try to destroy smoothness of $X_t(\cdot)$ and p tries to make $X_t(\cdot)$ smoother. Thus, it is natural to expect that $\{x : H_Z(x) = \eta\}$ can be described by jumps of a certain order depending on η.

Next we connect the martingale measure M with spectrally positive $1+\beta$-stable processes.

Let $L = \{L_t : t \ge 0\}$ denote a spectrally positive stable process of index $1+\beta$. Per definition, L is an $\mathbb{R}$-valued time-homogeneous process with independent increments and with Laplace transform given by

$$\mathbf{E}e^{-\lambda L_t} = e^{t\lambda^{1+\beta}}, \quad \lambda, t \ge 0. \tag{2.17}$$

Note that L is the unique (in law) solution to the following martingale problem:

$$t \mapsto e^{-\lambda L_t} - \int_0^t ds\, e^{-\lambda L_s} \lambda^{1+\beta} \text{ is a martingale for any } \lambda > 0. \tag{2.18}$$

In the next lemma we show that a stochastic integral of a non-negative function with respect to the martingale measure M can be represented as a time change of a spectrally positive $1 + \beta$ stable process. This representation simplifies significantly the analysis of regularity of super-Brownian motion with β-stable branching.

Lemma 2.2. *Let $d = 1$. Suppose $p \in (1 + \beta, 2)$ and let $\psi \in \mathscr{L}^p_{\mathrm{loc}}(\mu)$ with $\psi \geq 0$. Then there exists a spectrally positive $(1 + \beta)$-stable process $\{L_t : t \geq 0\}$ such that*

$$U_t(\psi) := \int_{(0,t] \times \mathbb{R}} M(\mathrm{d}(s,y)) \, \psi(s,y) = L_{T(t)}, \quad t \geq 0,$$

where $T(t) := \int_0^t \mathrm{d}s \int_{\mathbb{R}} X_s(\mathrm{d}y) \left(\psi(s,y) \right)^{1+\beta}$.

Proof. Let us write Itô's formula for $e^{-U_t(\psi)}$:

$$e^{-U_t(\psi)} - 1 = \text{local martingale}$$
$$+ \int_0^t \mathrm{d}s \, e^{-U_s(\psi)} \int_{\mathbb{R}} X_s(\mathrm{d}y) \int_0^\infty n(\mathrm{d}r) \left(e^{-r\psi(s,y)} - 1 + r\psi(s,y) \right).$$

Define $\tau(t) := T^{-1}(t)$ and put $t^* := \inf\{t : \tau(t) = \infty\}$. Then it is easy to get for every $v > 0$,

$$e^{-vU_{\tau(t)}(\psi)} = 1 + \int_0^t \mathrm{d}s \, e^{-vU_{\tau(s)}(\psi)} \frac{X_{\tau(s)} \left(v^{1+\beta} \psi^{1+\beta}(s, \cdot) \right)}{X_{\tau(s)} \left(\psi^{1+\beta}(s, \cdot) \right)} + \text{loc. mart.}$$
$$= 1 + \int_0^t \mathrm{d}s \, e^{-vU_{\tau(s)}(\psi)} v^{1+\beta} + \text{loc. mart.}, \quad t \leq t^*.$$

Since the local martingale is bounded, it is in fact a martingale. Let $\tilde{L}$ denote a spectrally positive process of index $1 + \beta$, independent of X. Define

$$L_t := \begin{cases} U_{\tau(t)}(\psi), & t \leq t^*, \\ U_{\tau(t^*)}(\psi) + \tilde{L}_{t-t^*}, & t > t^* \ (\text{if } t^* < \infty). \end{cases}$$

Then we can easily get that L satisfies the martingale problem (2.18) with κ replaced by $1 + \beta$. Now by time change back we obtain

$$U_t(\psi) = \tilde{L}_{T(t)} = L_{T(t)},$$

finishing the proof. $\qquad\qquad\qquad\qquad\qquad\qquad\qquad\qquad\qquad\qquad\qquad\qquad\qquad$ $\square$

Having this result we may represent the increment $Z_t(x_1) - Z_t(x_2)$ as a difference of two stable processes. More precisely, for every fixed pair (x_1, x_2) there exist spectrally positive stable processes L^+ and L^- such that

$$Z_t(x_1) - Z_t(x_2) = L^+_{T_+(x_1,x_2)} - L^-_{T_-(x_1,x_2)} \qquad\qquad (2.19)$$

where

$$T_{\pm}(x_1,x_2) = \int_0^t du \int_{\mathbb{R}} X_u(dy) \left((p_{t-u}(x_1-y) - p_{t-u}(x_2-y))^{\pm} \right)^{1+\beta}. \qquad (2.20)$$

It is clear from Lemma 2.2 that every jump $r\delta_{s,y}$ of the martingale measure M produces a jump of one of those stable processes:

- If $p_{t-s}(x_1-y) > p_{t-s}(x_2-y)$, then L^+ has a jump of size $r(p_{t-u}(x_1-y) - p_{t-u}(x_2-y))$;
- If $p_{t-s}(x_1-y) < p_{t-s}(x_2-y)$, then L^- has a jump of size $r(p_{t-u}(x_2-y) - p_{t-u}(x_1-y))$.

Therefore, representation (2.19) gives a handy tool for the study of the influence of jumps of M on the behavior of the increment $Z_t(x_1) - Z_t(x_2)$. Moreover, it becomes clear that one needs to know good estimates for the difference of kernels $p_{t-u}(x_1-y) - p_{t-u}(x_2-y)$ and for the tails of spectrally positive stable processes.

For Hölder exponents $\eta > 1$ we cannot use (2.19), since for exponents greater than 1 one has to subtract a polynomial correction. Instead of $Z_s(x_1) - Z_s(x_2)$, $s \leq t$, we shall consider

$$Z_s(x_1,x_2) := Z_s(x_1) - Z_s(x_2) - (x_1-x_2) \int_0^s \int_{\mathbb{R}} M(d(u,y)) \frac{\partial}{\partial y} p_{t-u}(x_2-y)$$

$$= \int_0^s \int_{\mathbb{R}} M(d(u,y)) q_{t-u}(x_1-y, x_2-y), \quad 0 \leq s \leq t, \qquad (2.21)$$

where

$$q_s(x,y) := p_s(x) - p_s(y) - (x-y) \frac{\partial}{\partial y} p_s(y). \qquad (2.22)$$

Here we may again apply Lemma 2.2 to obtain a representation for $Z_s(x_1,x_2)$, $s \leq t$ in terms of difference of spectrally positive stable processes, similarly to (2.19) which gives the representation for $Z_s(x_1) - Z_s(x_2)$. The only difference to (2.19) is that $p_{t-s}(x_1-y) - p_{t-s}(x_2-y)$ in (2.20) is replaced by $q_{t-s}(x_1-y, x_2-y)$.

Chapter 3
Some simple properties
of $(2, d, \beta)$-superprocesses

In this chapter we collect some estimates on $(2, d, \beta)$-superprocesses which are needed for the implementation of the program described in Chapter 1.

We start with a lemma where we give some left continuity properties of $(2, d, \beta)$-superprocess at fixed times, in dimensions $d < 2/\beta$.

Lemma 3.1. *Let $d < 2/\beta$, and B be an arbitrary open ball in $\mathbb{R}^d$. Then, for a fixed $t > 0$,*

$$\lim_{s \to t} X_s(B) = X_t(B), \quad \mathbf{P} - a.s.$$

Proof. Since t is fixed, X is continuous at t with probability 1. Therefore,

$$X_t(B) \leq \liminf_{s \to t} X_s(B) \leq \limsup_{s \to t} X_s(B) \leq \limsup_{s \to t} X_s(\overline{B}) \leq X_t(\overline{B})$$

with $\overline{B}$ denoting the closure of B. But since $X_t(\mathrm{d}x)$ is absolutely continuous with respect to Lebesgue measure, we have $X_t(B) = X_t(\overline{B})$. Thus the proof is finished. $\square$

In the next lemma we give a simple test for explosion of an integral involving $\{X_s(B)\}_{s \leq t}$ whereas B is an open ball in $\mathbb{R}^d$.

Lemma 3.2. *Let $d < 2/\beta$, and B be an arbitrary open ball in $\mathbb{R}^d$. Let $f : (0,t) \to (0,\infty)$ be measurable and assume that*

$$\int_{t-\delta}^{t} \mathrm{d}s\, f(t-s) = \infty \text{ for all sufficiently small } \delta \in (0,t).$$

Then for these δ

$$\int_{t-\delta}^{t} \mathrm{d}s\, X_s(B) f(t-s) = \infty \quad \textbf{P}\text{-a.s. on the event } \{X_t(B) > 0\}.$$

© The Author(s) 2016
L. Mytnik, V. Wachtel, *Regularity and Irregularity of Superprocesses with (1 + β)-stable Branching Mechanism*, SpringerBriefs in Probability and Mathematical Statistics, DOI 10.1007/978-3-319-50085-0_3

Proof. Fix δ as in the lemma. Fix also ω such that $X_t(B) > 0$ and $X_s(B) \to X_t(B)$ as $s \uparrow t$. For this ω, there is an $\varepsilon \in (0, \delta)$ such that $X_s(B) > \varepsilon$ for all $s \in (t - \varepsilon, t)$. Hence

$$\int_{t-\delta}^t ds\, X_s(B) f(t-s) \geq \varepsilon \int_{t-\varepsilon}^t ds\, f(t-s) = \infty,$$

and we are done.   □

Now we will study the properties of $(2, 1, \beta)$-superprocess in dimension $d = 1$. We start with moment estimates on the spatial increments of Z_t defined in (2.15). Until the end of this chapter we consider the case of

$$d = 1.$$

Lemma 3.3. *Let $d = 1$. For each $q \in (1, 1 + \beta)$ and $\delta < \min\{1, (2 - \beta)/(1 + \beta)\}$,*

$$\mathbf{E}|Z_t(x_1) - Z_t(x_2)|^q \leq C|x_1 - x_2|^{\delta q}, \quad x_1, x_2 \in \mathbb{R}.$$

Proof. Applying (3.1) from [32] with

$$\phi(s, y) = p_{t-s}(x_1 - y) - p_{t-s}(x_2 - y),$$

we get, for $\theta \in (1 + \beta, 2)$,

$$\mathbf{E}|Z_t(x_1) - Z_t(x_2)|^q$$
$$\leq C\left[\left(\int_0^t ds \int_{\mathbb{R}} S_s\mu(dy)|p_{t-s}(x_1 - y) - p_{t-s}(x_2 - y)|^\theta\right)^{q/\theta}\right.$$
$$\left. + \int_0^t ds \int_{\mathbb{R}} S_s\mu(dy)|p_{t-s}(x_1 - y) - p_{t-s}(x_2 - y)|^q\right]. \tag{3.1}$$

For every $\varepsilon \in (1, 3)$,

$$\int_0^t ds \int_{\mathbb{R}} S_s\mu(dy)|p_{t-s}(x_1 - y) - p_{t-s}(x_2 - y)|^\varepsilon$$
$$= \int_{\mathbb{R}} \mu(dz) \int_0^t ds \int_{\mathbb{R}} dy\, p_s(y - z)|p_{t-s}(x_1 - z) - p_{t-s}(x_2 - z)|^\varepsilon$$
$$= \int_{\mathbb{R}} \mu(dz) \int_0^t ds \int_{\mathbb{R}} dy\, p_s(y)|p_{t-s}(x_1 - z - y) - p_{t-s}(x_2 - z - y)|^\varepsilon.$$

Here and in many other places we need to control differences of Brownian transition kernels and their integrals. The corresponding bounds are derived in Appendix A. Now, using Lemma A.2 from that appendix, we get for every positive $\delta < \min\{1, (3 - \varepsilon)/\varepsilon\}$,

$$\int_0^t ds \int_{\mathbb{R}} S_s\mu(dy) \big| p_{t-s}(x_1 - y) - p_{t-s}(x_2 - y) \big|^\varepsilon$$

$$\leq C|x_1 - x_2|^{\delta\varepsilon} \int_{\mathbb{R}} \mu(dz) \Big(p_t\big((x_1 - z)/2\big) + p_t\big((x_2 - z)/2\big) \Big) \leq C|x_1 - x_2|^{\delta\varepsilon},$$

since μ, t are fixed. Applying this bound to both summands at the right-hand side of (3.1) finishes the proof of the lemma. □

Bounds on moments of spatial increments of Z_t, from the previous lemma, clearly give the same bounds on spatial increments of X_t itself. However, on top of this, they immediately give the bounds on the moments of the supremum of $X_t(\cdot)$ on compact spatial sets: this is done in the next lemma.

Lemma 3.4. *Let $d = 1$. If $K \subset \mathbb{R}$ is a compact and $q \in (1, 1+\beta)$, then*

$$\mathbf{E}\Big(\sup_{x \in K} X_t(x) \Big)^q < \infty.$$

Proof. By Jensen's inequality, we may additionally assume that $q > 1$. It follows from (2.14) that

$$\Big(\sup_{x \in K} X_t(x) \Big)^q \leq 4 \left(\Big(\sup_{x \in K} \mu * p_t(x) \Big)^q + \sup_{x \in K} \big| Z_t(x) \big|^q \right).$$

Clearly, the first term at the right-hand side is finite. Furthermore, according to Corollary 1.2 of Walsh [45], Lemma 3.3 implies that

$$\mathbf{E} \sup_{x \in K} \big| Z_t(x) \big|^q < \infty.$$

This completes the proof. □

From the above lemma one can immediately see that for any fixed t, $X_t(\cdot)$ is bounded on compacts. However, this is clearly not the case, if one start considering $X_s(x)$ as a function of (s, x) with $s \leq t$. The reason is obvious: as we have discussed in the introduction, the measure-valued process $X_s(dx)$ has jumps in the form of atomic measures, and if $\langle X_t, 1 \rangle > 0$, the set of jump times is dense in $[0, t]$. However, it turns out that if one "smooths" a bit X_s by taking its convolution with the heat kernel $p_{c(t-s)}(x - \cdot)$, then the resulting function of (s, x) is a.s. bounded on compacts for c large enough. This not obvious result is given in the next lemma.

Lemma 3.5. *Let $d = 1$. Fix a non-empty compact $K \subset \mathbb{R}$. Then*

$$V(K) := \sup_{0 \leq s \leq t, x \in K} S_{4(t-s)} X_s(x) < \infty \quad \mathbf{P} - a.s.$$

Proof. Assume that the statement of the lemma does not hold, i.e., there exists an event A of positive probability such that $\sup_{0 \leq s \leq t, x \in K} \overset{4}{S}_{4(t-s)} X_s(x) = \infty$ for every $\omega \in A$. Let $n \geq 1$. Put

$$\tau_n := \begin{cases} \inf\left\{ s < t : \text{ there exists } x \in K \text{ such that } S_{4(t-s)}X_s(x) > n \right\}, & \omega \in A, \\ t, & \omega \in A^c. \end{cases}$$

If $\omega \in A$, choose $x_n = x_n(\omega) \in K$ such that $S_{4(t-\tau_n)}X_{\tau_n}(x_n) > n$, whereas if $\omega \in A^c$, take any $x_n = x_n(\omega) \in K$. Using the strong Markov property gives

$$\mathbf{E}S_{3(t-\tau_n)}X_t(x_n) = \mathbf{E}\mathbf{E}\left[S_{3(t-\tau_n)}X_t(x_n) \,\middle|\, \mathscr{F}_{\tau_n} \right] \tag{3.2}$$

$$= \mathbf{E}S_{3(t-\tau_n)}S_{(t-\tau_n)}X_{\tau_n}(x_n) = \mathbf{E}S_{4(t-\tau_n)}X_{\tau_n}(x_n).$$

From the definition of (τ_n, x_n) we get

$$\mathbf{E}S_{4(t-\tau_n)}X_{\tau_n}(x_n) \geq n\mathbf{P}(A) \to \infty \text{ as } n \uparrow \infty.$$

In order to get a contradiction, we want to prove boundedness in n of the expectation in (3.2). Choosing a compact $K_1 \supset K$ satisfying $\text{dist}\big(K, (K_1)^c\big) \geq 1$, we have

$$\mathbf{E}S_{3(t-\tau_n)}X_t(x_n)$$

$$= \mathbf{E} \int_{K_1} dy\, X_t(y)\, p_{3(t-\tau_n)}(x_n - y) + \mathbf{E} \int_{(K_1)^c} dy\, X_t(y)\, p_{3(t-\tau_n)}(x_n - y)$$

$$\leq \mathbf{E} \sup_{y \in K_1} X_t(y) + \mathbf{E}X_t(\mathbb{R}) \sup_{y \in (K_1)^c, x \in K, 0 \leq s \leq t} p_{3s}(x - y).$$

By our choice of K_1 we obtain the bound

$$\mathbf{E}S_{3(t-\tau_n)}X_t(x_n) \leq \mathbf{E} \sup_{y \in K_1} X_t(y) + C = C, \tag{3.3}$$

the last step again by Lemma 3.4. Altogether, (3.2) is bounded in n, and the proof is finished.   □

It what follows let $B_r(x)$ denote an open ball with center x and radius r. Now, as an easy application of the previous lemma we have the following result:

Lemma 3.6. *Let $d = 1$. Fix any non-empty bounded $K \subset \mathbb{R}$. Then*

$$W_K := \sup_{(c,s,x):c \geq 1,\, 0 \vee (t-c^{-2}) \leq s < t, x \in K} \frac{X_s\big(B_{c(t-s)^{1/2}}(x)\big)}{c(t-s)^{1/2}} < \infty \quad \mathbf{P}-a.s.$$

Proof. Every ball of radius $c(t-s)^{1/2}$ can be covered with at most $\lceil c \rceil + 1$ balls of radius $(t-s)^{1/2}$. Therefore,

$$\sup_{(c,s,x):c \geq 1,\, 0 \vee (t-c^{-1/2}) \leq s < t, x \in K} \frac{X_s\big(B_{c(t-s)^{1/2}}(x)\big)}{c(t-s)^{1/2}}$$

$$\leq 2 \sup_{(s,x):0 < s \leq t, x \in K_1} \frac{X_s\big(B_{(t-s)^{1/2}}(x)\big)}{(t-s)^{1/2}},$$

where $K_1 := \{x : \text{dist}(x, \overline{K}) \leq 1\}$ with $\overline{K}$ denoting the closure of K. (The restriction $s \geq t - c^{-1/2}$ is imposed to have all centers x of the balls $B_{(t-s)^{1/2}}(x)$ in K_1.) We further note that

$$S_{t-s}X_s(x) = \int_{\mathbb{R}} dy \, p_{t-s}(x-y)X_s(y) \geq \int_{B_{(t-s)^{1/2}}(x)} dy \, p_{t-s}(x-y)X_s(y).$$

Using the monotonicity and the scaling property of p, we get the bound

$$S_{t-s}X_s(x) \geq (t-s)^{-1/2}p_1(1)X_s\big(B_{(t-s)^{1/2}}(x)\big).$$

Consequently,

$$\sup_{(s,x):0<s\leq t, x\in K_1} \frac{X_s\big(B_{(t-s)^{1/2}}(x)\big)}{(t-s)^{1/2}} \leq \frac{1}{p_1(1)} \sup_{(s,x):0<s\leq t, x\in K_1} S_{t-s}X_s(x).$$

It was proved in Lemma 3.5 that the random variable at the right-hand side is finite. Thus, the lemma is proved. $\qquad\square$

The boundedness of the smoothed density will play a crucial role in the analysis of the time changes $T_{\pm}(x_1,x_2)$ described in the previous chapter, see (2.19)–(2.22) and discussion there. The next lemma provides necessary tools to obtain pointwise upper bounds for $T_{\pm}(x_1,x_2)$: by taking $\theta = 1+\beta$ in (3.4) and (3.5) below we get estimates for $T_{\pm}(x_1,x_2)$.

Lemma 3.7. *Let $d = 1$. Fix $\theta \in [1,3)$, $\delta \in [0,1]$ with $\delta < (3-\theta)/\theta$, and a non-empty compact $K \subset \mathbb{R}$. Then*

$$\int_0^t ds \int_{\mathbb{R}} X_s(dy)\big|p_{t-s}(x_1-y) - p_{t-s}(x_2-y)\big|^{\theta}$$
$$\leq CV|x_1-x_2|^{\delta\theta}, \quad x_1,x_2 \in K, \quad \mathbf{P}-a.s., \tag{3.4}$$

with $V = V(K)$ from Lemma 3.5.
Moreover, for every $\theta \in [1,2)$ and $\delta \in (0,(3-2\theta)/\theta]$,

$$\int_0^t ds \int_{\mathbb{R}} X_s(dy)\Big|p_{t-s}(x_1-y) - p_{t-s}(x_2-y) - (x_1-x_2)\frac{\partial}{\partial x_2}p_{t-s}(x_2-y)\Big|^{\theta}$$
$$\leq CV|x_1-x_2|^{1+\delta}, \quad x_1,x_2 \in K, \quad \mathbf{P}-a.s. \tag{3.5}$$

Proof. Using (A.1) from Appendix A gives

$$\int_0^t ds \int_{\mathbb{R}} X_s(dy)\big|p_{t-s}(x_1-y) - p_{t-s}(x_2-y)\big|^{\theta} \leq C|x_1-x_2|^{\delta\theta} \times$$
$$\times \int_0^t ds\,(t-s)^{-(\delta\theta+\theta-1)/2} \int_{\mathbb{R}} X_s(dy)\Big(p_{t-s}\big((x_1-y)/2\big) + p_{t-s}\big((x_2-y)/2\big)\Big),$$

uniformly in $x_1, x_2 \in \mathbb{R}$. Recalling the scaling property of the kernel p, we get

$$\int_0^t ds \int_{\mathbb{R}} X_s(dy) |p_{t-s}(x_1 - y) - p_{t-s}(x_2 - y)|^\theta$$

$$\leq C|x_1 - x_2|^{\delta\theta} \int_0^t ds \, (t-s)^{-(\delta\theta + \theta - 1)/2} \left(S_{4(t-s)} X_s(x_1) + S_{4(t-s)} X_s(x_2) \right).$$

We complete the proof of (3.4) by applying Lemma 3.5. To derive (3.5) it suffices to replace (A.1) by (A.4) (from Appendix A) in the computations we used to prove (3.4).   □

In Lemmas 3.5 and 3.6 we have obtained uniform on compact sets upper bounds for the "smoothed" densities. Now we turn to the analysis of this smoothed density near a fixed spatial point. Without loss of generality we choose fixed point $x = 0$ in the next lemma.

Lemma 3.8. *Let $d = 1$. For all $c, \theta > 0$,*

$$\mathbf{P}\left(X_t(0) > \theta, \; \liminf_{s \uparrow t} S_{t-s} X_s \left(c \, (t-s)^{1/2} \right) \leq \theta \right) = 0.$$

Proof. For brevity, set

$$A := \left\{ \liminf_{s \uparrow t} S_{t-s} X_s \left(c \, (t-s)^{1/2} \right) \leq \theta \right\}$$

and for $n > 1/t$ define the stopping times

$$\tau_n := \begin{cases} \inf\left\{ s \in (t - 1/n, t) : S_{t-s} X_s \left(c \, (t-s)^{1/2} \right) \leq \theta + 1/n \right\}, & \omega \in A, \\ t, & \omega \in A^c. \end{cases}$$

Define also

$$x_n := c \, (t - \tau_n)^{1/2}.$$

Then, using the strong Markov property, we get

$$\mathbf{E}\left[X_t(x_n) \mid \mathscr{F}_{\tau_n} \right] = S_{t-\tau_n} X_{\tau_n}(x_n) = X_t(0) 1_{A^c} + S_{t-\tau_n} X_{\tau_n}(x_n) 1_A. \tag{3.6}$$

We next note that $x_n \to 0$ almost surely as $n \uparrow \infty$. This implies, in view of the continuity of X_t at zero, that $X_t(x_n) \to X_t(0)$ almost surely. Recalling that

$$\mathbf{E} \sup_{|x| \leq 1} X_t(x) < \infty,$$

in view of Corollary 2.8 of [18], we conclude that

$$X_t(x_n) \xrightarrow[n \uparrow \infty]{} X_t(0) \quad \text{in } \mathscr{L}_1.$$

This, in its turn, implies that

$$\mathbf{E}\left[X_t(x_n)|\mathscr{F}_{\tau_n}\right] - \mathbf{E}\left[X_t(0)|\mathscr{F}_{\tau_n}\right] \to 0 \quad \text{in } \mathscr{L}_1. \tag{3.7}$$

Furthermore, it follows from the well-known Lévy theorem on convergence of conditional expectations that

$$\mathbf{E}\left[X_t(0)\,\big|\,\mathscr{F}_{\tau_n}\right] \xrightarrow[n\uparrow\infty]{} \mathbf{E}\left[X_t(0)\,\big|\,\mathscr{F}_\infty\right] \quad \text{in } \mathscr{L}_1,$$

where $\mathscr{F}_\infty := \sigma\left(\cup_{n>1/t}\mathscr{F}_{\tau_n}\right)$.

Noting that $\tau_n \uparrow t$, we conclude that

$$\mathscr{F}_{t-} \subseteq \mathscr{F}_\infty \subseteq \mathscr{F}_t.$$

Since $X.(0)$ is continuous at fixed t a.s., we have $X_t(0) = \mathbf{E}\left[X_t(0)\,\big|\,\mathscr{F}_{t-}\right]$ almost surely. Consequently, $\mathbf{E}\left[X_t(0)\,\big|\,\mathscr{F}_\infty\right] = X_t(0)$ almost surely, and we get, as a result,

$$\mathbf{E}\left[X_t(0)\,\big|\,\mathscr{F}_{\tau_n}\right] \xrightarrow[n\uparrow\infty]{} X_t(0) \quad \text{in } \mathscr{L}_1. \tag{3.8}$$

Combining (3.7) and (3.8), we have

$$\mathbf{E}\left[X_t(x_n)\,\big|\,\mathscr{F}_{\tau_n}\right] \xrightarrow[n\uparrow\infty]{} X_t(0) \quad \text{in } \mathscr{L}_1.$$

From this convergence and from (3.6) we finally get

$$\mathbf{E}\left[1_A\,|X_t(0) - S_{t-\tau_n}X_{\tau_n}(x_n)|\right] \xrightarrow[n\uparrow\infty]{} 0.$$

Since $S_{t-\tau_n}X_{\tau_n}(x_n) \le \theta + 1/n$ on A, for all $n > 1/t$, the latter convergence implies that $X_t(0) \le \theta$ almost surely on the event A. Thus, the proof is finished. $\qquad\square$

Chapter 4
Analysis of jumps of superprocesses

This chapter is devoted to the analysis of jumps of $(2,d,\beta)$-superprocesses. The results of this chapter will be crucial for proofs of main theorems, since regularity properties of $(2,d,\beta)$-superprocesses depend heavily on presence and intensity of big jumps at certain locations. Note that the analysis of jumps of the processes is often very helpful in deriving their multifractal properties (see, e.g., [28] for the typical use of this method).

Let $\Delta X_s := X_s - X_{s-}$ denote the jumps of the measure-valued process X. Also let $|\Delta X_s| = \langle \Delta X_s, 1 \rangle$ be the size of the jump, and with some abuse of notation $|\Delta X_s(x)|$ denotes the size of jump at a point (s,x).

The results of the chapter are a bit technical, however let us explain briefly the main bounds we are going to obtain. Recall that $t > 0$ is fixed. First we would like to verify that the largest jump at the proximity of time t is of the order

$$|\Delta X_s| \sim (t-s)^{1/(1+\beta)}, \tag{4.1}$$

for $s < t$. The exact lower and upper bounds are given in Lemmas 4.1 and 4.2. Note that the jump of order $(t-s)^{1/(1+\beta)}$ happens at some "random" spatial point. Whenever one asks about the size of the maximal jump at the proximity of a given time-space point (t,x) one gets other estimates. For a moment fix a spatial point $x = 0$. It turns out that, if $X_t(0) > 0$, then the size of the maximal jump at the proximity of the time-space point $(t,0)$ is of order

$$|\Delta X_s(x)| \sim |(t-s)x|^{1/(1+\beta)}, \tag{4.2}$$

for x close to 0. This is shown in Lemma 4.3, Corollary 4.4, and Lemma 4.5.

We start with the lemma where we show that on any open subset of $\mathbb{R}^d$, "big" jumps will occur with probability one. In fact, in the lemma, we give a lower bound on the size of the largest jump.

© The Author(s) 2016
L. Mytnik, V. Wachtel, *Regularity and Irregularity of Superprocesses with (1 + β)-stable Branching Mechanism*, SpringerBriefs in Probability and Mathematical Statistics, DOI 10.1007/978-3-319-50085-0_4

Lemma 4.1. *Let $d < 2/\beta$ and B be an open ball in $\mathbb{R}^d$. For each $\varepsilon \in (0, t \wedge 1/2)$,*

$$\mathbf{P}\left(\Delta X_s(B) > (t-s)^{\frac{1}{1+\beta}} \log^{\frac{1}{1+\beta}}\left(\frac{1}{t-s}\right) \text{ for some } s \in (t-\varepsilon, t) \Big| X_t(B) > 0\right) = 1$$

Proof. It suffices to show that

$$\mathbf{P}\left(\Delta X_s(B) \leq (t-s)^{\frac{1}{1+\beta}} \log^{\frac{1}{1+\beta}}\left(\frac{1}{t-s}\right) \text{ for all } s \in (t-\varepsilon, t), X_t(B) > 0\right) = 0.$$
(4.3)

For $u \in (0, \varepsilon]$ define

$$\Pi_u := N\left((s, x, r) : s \in (t-\varepsilon, t-\varepsilon+u), x \in B, r > (t-s)^{\frac{1}{1+\beta}} \log^{\frac{1}{1+\beta}}\left(\frac{1}{t-s}\right)\right),$$

with the random measure N introduced in Lemma 2.1. Then

$$\left\{\Delta X_s(B) \leq (t-s)^{\frac{1}{1+\beta}} \log^{\frac{1}{1+\beta}}\left(\frac{1}{t-s}\right)\right\} = \{\Pi_\varepsilon = 0\}.$$
(4.4)

From a classical time change result for counting processes (see, e.g., Theorem 10.33 in Jacod [25]), we conclude that there exists a standard Poisson process $A = \{A(v) : v \geq 0\}$ such that

$$\Pi_u = A\left(\int_{t-\varepsilon}^{t-\varepsilon+u} ds\, X_s(B) \int_{(t-s)^{\frac{1}{1+\beta}} \log^{\frac{1}{1+\beta}}\left(\frac{1}{t-s}\right)}^{\infty} n(dr)\right)$$

$$= A\left(\frac{c_\beta}{1+\beta} \int_{t-\varepsilon}^{t-\varepsilon+u} ds\, X_s(B) \frac{1}{(t-s)\log\left(\frac{1}{t-s}\right)}\right),$$

where $c_\beta := \frac{\beta(\beta+1)}{\Gamma(1-\beta)}$. (By this definition, $n(dr) = c_\beta r^{-2-\beta} dr$.) Then

$$\mathbf{P}(\Pi_\varepsilon = 0, X_t(B) > 0) \leq \mathbf{P}\left(\int_{t-\varepsilon}^{t} ds\, X_s(B) \frac{1}{(t-s)\log\left(\frac{1}{t-s}\right)} < \infty, X_t(B) > 0\right).$$

It is easy to check that

$$\int_{t-\delta}^{t} ds\, \frac{1}{(t-s)\log\left(\frac{1}{t-s}\right)} = \infty \text{ for all } \delta \in (0, \varepsilon).$$

Therefore, by Lemma 3.2,

$$\int_{t-\varepsilon}^{t} ds\, X_s(B) \frac{1}{(t-s)\log\left(\frac{1}{t-s}\right)} = \infty \text{ on } \{X_t(B) > 0\}.$$

As a result we have $\mathbf{P}(\Pi_\varepsilon = 0, X_t(B) > 0) = 0$. Combining this with (4.4) we get (4.3). $\qquad\square$

The next result complements the previous lemma: it gives with probability close to one an upper bound for the sizes of jumps.

Lemma 4.2. *Let $d = 1$. Let $\varepsilon > 0$ and $\gamma \in (0, (1+\beta)^{-1})$. There exists a constant $c_{(4.5)} = c_{(4.5)}(\varepsilon, \gamma)$ such that*

$$\mathbf{P}\Big(|\Delta X_s| > c_{(4.5)} (t-s)^{(1+\beta)^{-1}-\gamma} \text{ for some } s < t \Big) \leq \varepsilon. \tag{4.5}$$

Proof. Recall the random measure N from Lemma 2.1(a). For any $c > 0$, set

$$Y_0 := N\Big([0, 2^{-1}t] \times \mathbb{R}^d \times (c 2^{-\lambda}t^\lambda, \infty)\Big), \tag{4.6}$$

$$Y_n := N\Big([(1 - 2^{-n})t, (1 - 2^{-n-1})t) \times \mathbb{R}^d \times (c 2^{-\lambda(n+1)}t^\lambda, \infty)\Big), \quad n \geq 1, \tag{4.7}$$

where $\lambda := (1+\beta)^{-1} - \gamma$. It is easy to see that

$$\mathbf{P}\Big(|\Delta X_s| > c (t-s)^\lambda \text{ for some } s < t \Big) \leq \mathbf{P}\Big(\sum_{n=0}^\infty Y_n \geq 1 \Big) \leq \sum_{n=0}^\infty \mathbf{E} Y_n, \tag{4.8}$$

where in the last step we have used the classical Markov inequality. From the formula for the compensator $\hat{N}$ of N in Lemma 2.1(b),

$$\mathbf{E} Y_n = c_\beta \int_{(1-2^{-n})t}^{(1-2^{-n-1})t} ds\, \mathbf{E} X_s(\mathbb{R}^d) \int_{c 2^{-\lambda(n+1)}t^\lambda}^\infty dr\, r^{-2-\beta}, \quad n \geq 1. \tag{4.9}$$

Now

$$\mathbf{E} X_s(\mathbb{R}^d) = X_0(\mathbb{R}^d) =: c_{(4.10)}. \tag{4.10}$$

Consequently,

$$\mathbf{E} Y_n \leq \frac{c_\beta}{1+\beta} c_{(4.10)} c^{-1-\beta} 2^{-(n+1)\gamma(1+\beta)} t^{\gamma(1+\beta)}. \tag{4.11}$$

Analogous calculations show that (4.11) remains valid also in the case $n = 0$. Therefore,

$$\sum_{n=0}^\infty \mathbf{E} Y_n \leq \frac{c_\beta}{1+\beta} c_{(4.10)} c^{-1-\beta} t^{\gamma(1+\beta)} \sum_{n=0}^\infty 2^{-(n+1)\gamma(1+\beta)}$$

$$= \frac{c_\beta}{1+\beta} c_{(4.10)} c^{-1-\beta} t^{\gamma(1+\beta)} \frac{2^{-\gamma(1+\beta)}}{1 - 2^{-\gamma(1+\beta)}}. \tag{4.12}$$

Choosing $c = c_{(4.5)}$ such that the expression in (4.12) equals ε, and combining with (4.8), the proof is complete. $\qquad\square$

Put

$$f_{s,x} := \log\big(1 + (t-s)^{-1}\big) \log\big(1 + |x|^{-1}\big) 1_{\{x \neq 0\}} 1_{\{s < t\}}. \tag{4.13}$$

In the following lemma and corollary we obtain suitable upper bounds for maximal jumps which occur close to 0.

Lemma 4.3. *Let $d = 1$. Fix $X_0 = \mu \in \mathcal{M}_f \backslash \{0\}$. Let $\varepsilon > 0$ and $q > 0$. Then there exists a constant $c_{(4.14)} = c_{(4.14)}(\varepsilon, q)$ such that*

$$\mathbf{P}\left(\Delta X_s(x) > c_{(4.14)}\left((t-s)|x|\right)^{\frac{1}{1+\beta}} (f_{s,x})^\ell \text{ for some } s < t, \ x \in B_2(0)\right) \leq \varepsilon, \quad (4.14)$$

where

$$\ell := \frac{1}{1+\beta} + q. \quad (4.15)$$

Proof. For any $c > 0$ (later to be specialized to some $c_{(4.19)}$) set

$$Y := N\left((s,x,r) : (s,x) \in (0,t) \times B_2(0), \ r \geq c\left((t-s)|x|\right)^{1/(1+\beta)} (f_{s,x})^\ell\right),$$

Clearly,

$$\mathbf{P}\left(\Delta X_s(x) > c\left((t-s)|x|\right)^{1/(1+\beta)} (f_{s,x})^\ell \text{ for some } s < t \text{ and } x \in B_2(0)\right)$$
$$= \mathbf{P}(Y \geq 1) \leq \mathbf{E}Y, \quad (4.16)$$

where in the last step we have used the classical Markov inequality. From (2.3),

$$\mathbf{E}Y = c_\beta \, \mathbf{E} \int_0^t ds \int_{\mathbb{R}} X_s(dx) \, 1_{B_2(0)}(x) \int_{c\left((t-s)|x|\right)^{1/(1+\beta)}(f_{s,x})^\ell}^\infty dr \, r^{-2-\beta}$$

$$= c_\beta \frac{c^{-1-\beta}}{1+\beta} \int_0^t ds \, (t-s)^{-1} \left(\log\left(1 + (t-s)^{-1}\right)\right)^{-1-q(1+\beta)} \quad (4.17)$$

$$\times \int_{\mathbb{R}} \mathbf{E}X_s(dx) \, 1_{B_2(0)}(x) \, |x|^{-1} \left(\log\left(1 + |x|^{-1}\right)\right)^{-1-q(1+\beta)}.$$

Now, writing C for a generic constant (which may change from place to place),

$$\int_{\mathbb{R}} \mathbf{E}X_s(dx) \, 1_{B_2(0)}(x) \, |x|^{-1} \left(\log\left(1 + |x|^{-1}\right)\right)^{-1-q(1+\beta)}$$

$$\leq \int_{\mathbb{R}} \mu(dy) \int_{\mathbb{R}} dx \, p_s(x-y) \, 1_{B_2(0)}(x) \, |x|^{-1} \left(\log\left(1 + |x|^{-1}\right)\right)^{-1-q(1+\beta)}$$

$$\leq C\mu(\mathbb{R}) s^{-1/2} \int_{\mathbb{R}} dx \, 1_{B_2(0)}(x) \, |x|^{-1} \left(\log\left(1 + |x|^{-1}\right)\right)^{-1-q(1+\beta)}$$

$$=: c_{(4.18)} s^{-1/2}, \quad (4.18)$$

where $c_{(4.18)} = c_{(4.18)}(q)$ (recall that t is fixed). Consequently,

$$\mathbf{E}Y \leq c_\beta \, c_{(4.18)} \, c^{-1-\beta} \int_0^t ds \, s^{-1/2} \, (t-s)^{-1} \left(\log\left(1 + (t-s)^{-1}\right)\right)^{-1-q(1+\beta)}$$

$$=: c_{(4.19)} \, c^{-1-\beta} \quad (4.19)$$

with $c_{(4.19)} = c_{(4.19)}(q)$. Choose now c such that the latter expression equals ε and write $c_{(4.19)}$ instead of c. Recalling (4.16), the proof is complete.

Since $\sup_{0<y<1} y^{\gamma} \log^{\ell} \frac{1}{y} < \infty$ for every $\gamma > 0$, the following corollary is immediate from Lemma 4.3:

Corollary 4.4. *Fix* $X_0 = \mu \in \mathscr{M}_{\mathrm{f}} \backslash \{0\}$. *Let* $d = 1$. *Let* $\varepsilon > 0$ *and* $\gamma \in \left(0, (1+\beta)^{-1}\right)$. *There exists a constant* $c_{(4.20)} = c_{(4.20)}(\varepsilon, \gamma)$ *such that*

$$\left\{ \Delta X_s(x) > c_{(4.20)} \left((t-s)|x|\right)^{\frac{1}{1+\beta} - \gamma} \text{ for some } s < t \text{ and } x \in B_2(0) \right\}$$

$$\subset \left\{ \Delta X_s(x) > c_{(4.14)} \left((t-s)|x|\right)^{\frac{1}{1+\beta}} (f_{s,x})^{\ell} \text{ for some } s < t, \ x \in B_2(0) \right\}. \quad (4.20)$$

In particular,

$$\mathbf{P}\left(\Delta X_s(x) > c_{(4.20)} \left((t-s)|x|\right)^{\frac{1}{1+\beta} - \gamma} \text{ for some } s < t \text{ and } x \in B_2(0) \right) \leq \varepsilon.$$

Introduce the event

$$D_\theta := \left\{ X_t(0) > \theta, \ \sup_{0 < s \leq t} X_s(\mathbb{R}) \leq \theta^{-1}, \ W_{B_3}(0) \leq \theta^{-1} \right\}. \quad (4.21)$$

In the next lemma we obtain a lower bound on a jump which occurs close to time t and to the spatial point $x_0 = 0$.

Lemma 4.5. *Let* $d = 1$. *For each* $\theta > 0$ *there exists a constant* $c_{(4.22)} = c_{(4.22)}(\theta) \geq 1$ *such that*

$$\mathbf{P}\left(\Delta X_s(y) > Q \left(y(t-s)\right)^{1/(1+\beta)} \log^{1/(1+\beta)} \left((t-s)^{-1}\right) \right) \quad (4.22)$$

for some $s \in (t - \varepsilon, t)$ *and* $\dfrac{c_{(4.22)}}{2} (t-s)^{1/2} \leq y \leq \dfrac{3c_{(4.22)}}{2} (t-s)^{1/2} \Big| D_\theta \Big) = 1$

for all $\varepsilon \in (0, t \wedge 1/8)$, $Q > 0$.

Proof. Analogously to the proof of Lemma 4.1, to show that the number of jumps is greater than zero almost surely on some event, it is enough to show the divergence of a certain integral on that event or even on a bigger one. Specifically here, it suffices to verify that there exists $c = c_{4.22}$ such that

$$I_{\varepsilon,c} := \int_{t-\varepsilon}^{t} \frac{ds}{(t-s)\log\left((t-s)^{-1}\right)} \int_{\frac{c}{2}(t-s)^{1/2}}^{\frac{3c}{2}(t-s)^{1/2}} dy \, y^{-1} X_s(y) = \infty$$

almost surely on the event D_θ.

The mapping $\varepsilon \mapsto I_{\varepsilon,c}$ is nonincreasing. Therefore, we shall additionally assume, without loss of generality, that $\varepsilon \leq c^{-1/2}$ and this in turn implies that $c(t-s)^{1/2} \leq 1$ for all $s \in (t - \varepsilon, t)$. So, in what follows, in the proof of the lemma we will assume without loss of generality that given c, we choose ε so that

$$c(t-s)^{1/2} \leq 1, \quad \forall s \in (t - \varepsilon, t).$$

Since $y \le \frac{3c}{2}(t-s)^{1/2}$ and $p_s(x) \le p_s(0)$ for all $x \in \mathbb{R}$, we have

$$I_{\varepsilon,c} \ge \frac{2}{3c} \int_{t-\varepsilon}^{t} \frac{ds}{(t-s)^{3/2} \log((t-s)^{-1})} \\ \times \int_{\frac{c}{2}(t-s)^{1/2}}^{\frac{3c}{2}(t-s)^{1/2}} dy \, \frac{p_{t-s}(c(t-s)^{1/2}-y)}{p_{t-s}(0)} X_s(y).$$

Then, using the scaling property of the kernel p, we obtain

$$I_{\varepsilon,c} \ge \frac{2}{3c \, p_1(0)} \int_{t-\varepsilon}^{t} \frac{ds}{(t-s) \log((t-s)^{-1})} \left(S_{t-s} X_s \left(c(t-s)^{1/2} \right) \right. \\ \left. - \int_{|y-c(t-s)^{1/2}| > \frac{c}{2}(t-s)^{1/2}} dy \, p_{t-s}\left(c(t-s)^{1/2} - y \right) X_s(y) \right). \quad (4.23)$$

Since we are in dimension one, if

$$y \in \widetilde{D}_{s,j} := \left\{ z : c\left(\frac{1}{2} + j \right)(t-s)^{1/2} < \left| z - c(t-s)^{1/2} \right| \right. \\ \left. < c\left(2 + \frac{1}{2} + j \right)(t-s)^{1/2} \right\}, \quad (4.24)$$

then

$$p_{t-s}\left(c(t-s)^{1/2} - y \right) \le p_{t-s}\left(c(j+1/2)(t-s)^{1/2} \right) \\ = (t-s)^{-1/2} p_1\left(c(j+1/2) \right) = (2\pi)^{-1/2}(t-s)^{-1/2} e^{-c^2(1/2+j)^2/2}.$$

From this bound we conclude that

$$\int_{|y-c(t-s)^{1/2}| > \frac{c}{2}(t-s)^{1/2}} dy \, p_{t-s}\left(c(t-s)^{1/2} - y \right) 1_{B_2(0)}(y) X_s(y) \\ \le (2\pi)^{-1/2}(t-s)^{-1/2} \sum_{j=0}^{\infty} e^{-c^2(1/2+j)^2/2} \int_{\widetilde{D}_{s,j}} dy 1_{B_2(0)}(y) X_s(y).$$

Now recall again that the spatial dimension equals to one and hence for any $j \ge 0$ the set $\widetilde{D}_{s,j}$ in (4.24) is the union of two balls of radius $c(t-s)^{1/2}$. If furthermore $\widetilde{D}_{s,j} \cap B_2(0) \ne \emptyset$, then, in view of the assumption $c(t-s)^{1/2} \le 1$, the centers of those balls lie in $B_3(0)$. Therefore, we can apply Lemma 3.6 to bound the integral $\int_{\widetilde{D}_{s,j}} dy 1_{B_2(0)}(y) X_s(y)$ by $2c(t-s)^{1/2} W_{B_3(0)}$ and obtain

$$\int_{|y-c(t-s)^{1/2}| > \frac{c}{2}(t-s)^{1/2}} dy \, p_{t-s}\left(c(t-s)^{1/2} - y \right) 1_{B_2(0)}(y) X_s(y) \\ \le \frac{2c}{(2\pi)^{1/2}} W_{B_3(0)} \sum_{j=0}^{\infty} e^{-c^2(1/2+j)^2/2} \le C W_{B_3(0)} c^{-2}. \quad (4.25)$$

Furthermore, if $|y| \geq 2$ and $(t-s) \leq c^{-2}$, then

$$p_{t-s}\big(c\,(t-s)^{1/2} - y\big) \leq p_{t-s}(1) = (t-s)^{-1/2} p_1\big((t-s)^{-1/2}\big)$$
$$= (2\pi)^{-1/2}(t-s)^{-1/2}e^{-1/2(t-s)}.$$

This implies that

$$\int_{\mathbb{R}\backslash B_2(0)} dy \, p_{t-s}\big(c\,(t-s)^{1/2} - y\big)X_s(y) \leq (2\pi)^{-1/2}(t-s)^{-1/2}e^{-1/2(t-s)}X_s(\mathbb{R})$$
$$\leq Cc^{-2}X_s(\mathbb{R}).$$

Combining this bound with (4.25), we obtain

$$\int_{\left|y-c(t-s)^{1/2}\right| > \frac{c}{2}(t-s)^{1/2}} dy \, p_{t-s}\big(c\,(t-s)^{1/2} - y\big)X_s(y)$$
$$\leq Cc^{-2}\Big(W_{B_3(0)} + \sup_{0<s\leq t} X_s(\mathbb{R})\Big).$$

Thus, we can choose c so large that the right-hand side in the previous inequality does not exceed $\theta/2$. Since, in view of Lemma 3.8,

$$\liminf_{s\uparrow t} S_{t-s}X_s\big(c\,(t-s)^{1/2}\big) > \theta,$$

we finally get

$$\liminf_{s\uparrow t}\Bigg(S_{t-s}X_s\big(c\,(t-s)^{1/2}\big)$$
$$- \int_{\left|y-c(t-s)^{1/2}\right| > \frac{c}{2}(t-s)^{1/2}} dy \, p_{t-s}\big(c\,(t-s)^{1/2} - y\big)X_s(y)\Bigg) \geq \theta/2.$$

From this bound and (4.23) the desired property of $I_{\varepsilon,c}$ follows. $\square$

Fix any $\theta > 0$, and to simplify notation write $c := c_{(4.22)}$. For all n sufficiently large, say $n \geq N_0$, define

$$A_n := \Big\{\Delta X_s\Big(\big(\tfrac{c}{2}2^{-n}, \tfrac{3c}{2}2^{-n}\big)\Big) \geq 2^{-(\bar{n}_c+1)n}\, n^{1/(1+\beta)}$$
$$\text{for some } s \in \big(t-2^{-2n}, t-2^{-2(n+1)}\big)\Big\}. \quad (4.26)$$

Based on Lemma 4.5 we will show in the following lemma that if $X_t(0) > 0$ then there exist infinitely many jumps $\Delta X_s(x)$ which are greater than $((t-s)|x|)^{1/(1+\beta)}$ with $x \sim (t-s)^{1/2}$. To be more precise, we show that A_n occur infinitely often.

Lemma 4.6. *We have*

$$\mathbf{P}\big(A_n \text{ infinitely often} \,\big|\, D_\theta\big) = 1.$$

Proof. If $y \in \big(\frac{\varsigma}{2}(t-s)^{1/2}, \frac{3c}{2}(t-s)^{1/2}\big)$ and $s \in (t - 2^{-2n}, t - 2^{-2(n+1)})$, then

$$\big((t-s)y\log\big((t-s)^{-1}\big)\big)^{1/(1+\beta)} \geq \Big(2^{-2(n+1)}\frac{c}{2}2^{-n-1}2n\log 2\Big)^{1/(1+\beta)}$$

$$= c_{(4.27)}^{-1}2^{-(\tilde{n}_c+1)n}n^{1/(1+\beta)}. \qquad (4.27)$$

This implies that

$$A_n \supseteq \Big\{\Delta X_s(y) \geq c_{(4.27)}\big((t-s)y\log\big((t-s)^{-1}\big)\big)^{1/(1+\beta)} \qquad (4.28)$$

$$\text{for some } s \in (t - 2^{-2n}, t - 2^{-2(n+1)}) \text{ and } y \in \Big(\frac{c}{2}(t-s)^{1/2}, \frac{3c}{2}(t-s)^{1/2}\Big)\Big\}.$$

Consequently, from (4.28) we get

$$\bigcup_{n=N}^{\infty} A_n \supseteq \Big\{\Delta X_s(y) \geq c_{(4.27)}\big((t-s)y\log\big((t-s)^{-1}\big)\big)^{1/(1+\beta)}$$

$$\text{for some } s \in (t - 2^{-2N}, t) \text{ and } y \in \Big(\frac{c}{2}(t-s)^{1/2}, \frac{3c}{2}(t-s)^{1/2}\Big)\Big\}$$

for all $N > N_0 \vee \frac{1}{2}\log_2(t \wedge 1/8)$. Applying Lemma 4.5 and using the monotonicity of the union in N, we get

$$\mathbf{P}\Big(\bigcup_{n=N}^{\infty} A_n \,\Big|\, D_\theta\Big) = 1 \quad \text{for all } N \geq N_0.$$

This completes the proof. $\qquad \square$

Chapter 5
Dichotomy for densities

5.1 Proof of Theorem 1.1(a)

The non-random part $\mu * p_t(x)$ is differentiable. The continuity of $Z_t(\cdot)$ follows from the classical

Kolmogorov criterion. *If stochastic process $\xi(t)$ satisfies*

$$\mathbf{E}|\xi(t) - \xi(s)|^a \leq C|t - s|^{1+b}$$

for some $a, b > 0$, then there exists a Hölder continuous version with any index η smaller than b/a.

Combing Lemma 3.3 with the Kolmogorov criterion with $a = q$, $b = \delta q - 1$, we infer that $Z_t(\cdot)$ is Hölder continuous of all orders smaller than $\delta - 1/q$ if we can choose δ and q such that $\delta q > 1$. Noting that

$$\sup_{\delta < \min\{1, (2-\beta)/(1+\beta)\}, q \in (1, 1+\beta)} (\delta - 1/q) = \min\left\{1, \frac{2-\beta}{1+\beta}\right\} - \frac{1}{1+\beta} > 0$$

for all $\beta \in (0,1)$. Hence, we see that $Z_t(\cdot)$ is Hölder continuous of all orders smaller than $\min\{\beta, 1 - \beta\}/(1+\beta)$. In other words, we proved Theorem 1.2 for $\beta \geq 1/2$.

5.2 Proof of Theorem 1.1(b)

Throughout this section we assume that $d > 1$. Recall that $t > 0$ and $X_0 = \mu \in \mathcal{M}_f \setminus \{0\}$ are fixed. We want to verify that for each version of the density function X_t the property

$$\|X_t\|_B = \infty \quad \mathbf{P}\text{-a.s. on the event } \{X_t(B) > 0\} \tag{5.1}$$

© The Author(s) 2016
L. Mytnik, V. Wachtel, *Regularity and Irregularity of Superprocesses with (1 + β)-stable Branching Mechanism*, SpringerBriefs in Probability and Mathematical Statistics, DOI 10.1007/978-3-319-50085-0_5

holds whenever B is a fixed open ball in $\mathbb{R}^d$. Having this relation for every open ball we may prove Theorem 1.1(b) by the following simple argument: Let fix ω outside a null set so that (5.1) is valid for any ball with rational center and rational radius. If U is an open set with $X_t(U) > 0$, then there exists a ball B with rational center and rational radius such that $B \subset U$ and $X_t(B) > 0$. Consequently, $\|X_t\|_U(\omega) = \|X_t\|_B(\omega) = \infty$.

To get (5.1) we first show that on the event $\{X_t(B) > 0\}$ there are always sufficiently "big" jumps of X on B that occur close to time t. This is done in Lemma 4.1 above. Then with the help of properties of the log-Laplace equation derived in Lemma 5.1 we are able to show that the "big" jumps are large enough to ensure the unboundedness of the density at time t. Loosely speaking the density is getting unbounded in the proximity of big jumps. As we have seen in the previous chapter, the largest jump at time $s < t$ is of order $(t - s)^{1/(1+\beta)}$. Suppose this jump occurs at spatial point x. Since a jump occurring at time s is smeared out by the kernel p_{t-s}, we have the following estimate for the value of the density at time t and spatial point x:

$$X_t(x) \approx (t-s)^{1/(1+\beta)} p_{t-s}(0) \approx (t-s)^{1/(1+\beta)-d/2}. \tag{5.2}$$

From (5.2) it is clear that the density should explode in any dimension $d > 1$. In the rest of the chapter we justify this heuristic.

Set $\varepsilon_n := 2^{-n}$, $n \geq 1$. Then we choose open balls $\mathbf{B}_n \uparrow B$ such that

$$\overline{\mathbf{B}_n} \subset \mathbf{B}_{n+1} \subset B \quad \text{and} \quad \sup_{y \in B^c, x \in \mathbf{B}_n, 0 < s \leq \varepsilon_n} p_s(x-y) \xrightarrow[n \uparrow \infty]{} 0. \tag{5.3}$$

Fix $n \geq 1$ such that $\varepsilon_n < t$. Set, for brevity,

$$\tau_n := \inf \left\{ s \in (t - \varepsilon_n, t) : \Delta X_s(\mathbf{B}_n) > (t-s)^{\frac{1}{1+\beta}} \log^{\frac{1}{1+\beta}} \left(\frac{1}{t-s} \right) \right\}.$$

It follows from Lemma 4.1 that

$$\mathbf{P}(\tau_n = \infty) \leq \mathbf{P}(X(\mathbf{B}_n) = 0), \quad n \geq 1. \tag{5.4}$$

In order to obtain a lower bound for $\|X_t\|_B$ we use the following inequality:

$$\|X_t\|_B \geq \int_B dy\, X_t(y) p_u(y - x), \quad x \in B,\ u > 0. \tag{5.5}$$

On the event $\{\tau_n < t\}$, denote by ζ_n the spatial location in $\mathbf{B}_n$ of the jump at time τ_n, and by r_n the size of the jump, meaning that $\Delta X_{\tau_n} = r_n \delta_{\zeta_n}$. Then specializing (5.5),

$$\|X_t\|_B \geq \int_B dy\, X_t(y)\, p_{t-\tau_n}(y - \zeta_n) \text{ on the event } \{\tau_n < t\}. \tag{5.6}$$

From the strong Markov property at time τ_n, together with the branching property of superprocesses, we know that conditionally on $\{\tau_n < t\}$, the process $\{X_{\tau_n+u} : u \geq 0\}$

is bounded below in distribution by $\{\widetilde{X}_u^n : u \geq 0\}$, where $\widetilde{X}^n$ is a super-Brownian motion with initial value $r_n \delta_{\zeta_n}$. Hence, from (5.6) we get

$$\mathbf{E}\exp\{-\|X_t\|_B\} \tag{5.7}$$

$$\leq \mathbf{E}1_{\{\tau_n < t\}} \exp\left\{-\int_B dy\, X_t(y)\, p_{t-\tau_n}(y - \zeta_n)\right\} + \mathbf{P}(\tau_n = \infty)$$

$$\leq \mathbf{E}1_{\{\tau_n < t\}} \mathbf{E}_{r_n \delta_{\zeta_n}} \exp\left\{-\int_B dy\, X_{t-\tau_n}(y)\, p_{t-\tau_n}(y - \zeta_n)\right\} + \mathbf{P}(\tau_n = \infty).$$

Note that on the event $\{\tau_n < t\}$, we have

$$r_n \geq (t - \tau_n)^{\frac{1}{1+\beta}} \log^{\frac{1}{1+\beta}}\left(\frac{1}{t - \tau_n}\right) =: h_\beta(t - \tau_n). \tag{5.8}$$

We now claim that

$$\lim_{n\uparrow\infty} \sup_{0 < s < \varepsilon_n,\, x \in \mathbf{B}_n,\, r \geq h_\beta(s)} \mathbf{E}_{r\delta_x} \exp\left\{-\int_B dy\, X_s(y) p_s(y - x)\right\} = 0. \tag{5.9}$$

To verify (5.9), let $s \in (0, \varepsilon_n)$, $x \in \mathbf{B}_n$, and $r \geq h_\beta(s)$. Then, using the Laplace transition functional of the superprocess we get

$$\mathbf{E}_{r\delta_x} \exp\left\{-\int_B dy\, X_s(y) p_s(y - x)\right\} = \exp\left\{-r v_{s,x}^n(s, x)\right\}$$

$$\leq \exp\left\{-h_\beta(s) v_{s,x}^n(s, x)\right\}, \tag{5.10}$$

where the non-negative function $v_{s,x}^n = \{v_{s,x}^n(s', x') : s' > 0,\, x' \in \mathbb{R}^d\}$ solves the log-Laplace integral equation

$$v_{s,x}^n(s', x') = \int_{\mathbb{R}^d} dy\, p_{s'}(y - x')\, 1_B(y)\, p_s(y - x) \tag{5.11}$$

$$- \int_0^{s'} dr' \int_{\mathbb{R}^d} dy\, p_{s'-r'}(y - x') \left(v_{s,x}^n(r', y)\right)^{1+\beta}$$

related to (1.6).

Lemma 5.1. *If* $d > 1$, *then*

$$\lim_{n\uparrow\infty}\left(\inf_{0 < s < \varepsilon_n,\, x \in \mathbf{B}_n} h_\beta(s) v_{s,x}^n(s, x)\right) = +\infty. \tag{5.12}$$

Proof. We start with a determination of the asymptotics of the first term at the right-hand side of the log-Laplace equation (5.11) at $(s', x') = (s, x)$. Note that

$$\int_{\mathbb{R}^d} dy\, p_s(y - x)\, 1_B(y)\, p_s(y - x) \tag{5.13}$$

$$= \int_{\mathbb{R}^d} dy\, p_s(y - x)\, p_s(y - x) - \int_{B^c} dy\, p_s(y - x)\, p_s(y - x).$$

In the latter formula line, the first term equals $p_{2s}(0) = Cs^{-d/2}$, whereas the second one is bounded from above by

$$\sup_{0<s<\varepsilon_n, x\in \mathbf{B}_n, \, y\in B^c} p_s(y-x) \xrightarrow[n\uparrow\infty]{} 0, \tag{5.14}$$

where the last convergence follows by assumption (5.3) on $\mathbf{B}_n$. Hence from (5.13) and (5.14) we obtain

$$\int_{\mathbb{R}^d} dy \, p_s(y-x) 1_B(y) \, p_s(y-x) = Cs^{-d/2} + o(1) \text{ as } n\uparrow\infty, \tag{5.15}$$

uniformly in $s \in (0, \varepsilon_n)$ and $x \in \mathbf{B}_n$.

To simplify notation, we write $v^n := v^n_{s,x}$. Next, since v^n is non-negative we drop the non-linear term in (5.11) to get the upper bound

$$v^n(s', x') \leq \int_{\mathbb{R}^d} dy \, p_{s'}(y-x') p_s(y-x) = p_{s'+s}(x-x').$$

Then we have

$$\int_0^s dr' \int_{\mathbb{R}^d} dy \, p_{s-r'}(y-x)\left(v^n(r', y)\right)^{1+\beta} \tag{5.16}$$

$$\leq \int_0^s dr' \int_{\mathbb{R}^d} dy \, p_{s-r'}(y-x)\left(p_{r'+s}(x-y)\right)^{1+\beta}$$

$$\leq \left(p_s(0)\right)^\beta \int_0^s dr' \int_{\mathbb{R}^d} dy \, p_{s-r'}(y-x) \, p_{r'+s}(x-y)$$

$$= \left(p_s(0)\right)^\beta \int_0^s dr' \, p_{2s}(0) = Cs^{1-d(1+\beta)/2}.$$

Summarizing, by (5.11), (5.15), and (5.16),

$$v^n(s,x) \geq Cs^{-d/2} + o(1) - Cs^{1-d(1+\beta)/2} \tag{5.17}$$

uniformly in $s \in (0, \varepsilon_n)$ and $x \in \mathbf{B}n$. According to the general assumption $d < 2/\beta$, we conclude that the right-hand side of (5.17) behaves like $Cs^{-d/2}$ as $s \downarrow 0$, uniformly in $s \in (0, \varepsilon_n)$. Now recalling definition (5.8) as well as our assumption that $d > 1$ we immediately get

$$\lim_{n\uparrow\infty} \inf_{0<s<\varepsilon_n} h_\beta(s) s^{-d/2} = +\infty.$$

By (5.17), this implies (5.12), and the proof of the lemma is finished. $\qquad\square$

We are now in position to complete the proof of Theorem 1.1(b). The claim (5.9) readily follows from estimate (5.10) and (5.12). Moreover, according to (5.9), by passing to the limit $n\uparrow\infty$ in the right-hand side of (5.7), and then using (5.4), we arrive at

$$\mathbf{E}\exp\left\{-\|X_t\|_B\right\} \leq \limsup_{n\uparrow\infty} \mathbf{P}(\tau_n = \infty) \leq \limsup_{n\uparrow\infty} \mathbf{P}(X_t(\mathbf{B}_n) = 0).$$

Since the event $\{X_t(B) = 0\}$ is the nonincreasing limit as $n \uparrow \infty$ of the events $\{X_t(\mathbf{B}_n) = 0\}$ we get

$$\mathbf{E}\exp\{ - \|X_t\|_B\} \leq \mathbf{P}(X_t(B) = 0).$$

Since obviously $\|X_t\|_B = 0$ if and only if $X_t(B) = 0$, we see that (5.1) follows from this last bound. The proof of Theorem 1.1(b) is finished for $U = B$.

Chapter 6
Pointwise Hölder exponent at a given point: proof of Theorem 1.3

Let us first give a heuristic explanation for the value of $\bar{\eta}_c$. According to Lemmas 4.3 and 4.5, the maximal jump at time s and spatial point x near point $x_0 = 0$ is of order $((t - s)|x|)^{1/(1+\beta)}$. Due to the scaling properties of the heat kernel, the jump that has a decisive effect on the pointwise Hölder exponent at $x_0 = 0$ should occur at distance

$$|x| \approx (t - s)^{1/2}. \tag{6.1}$$

Then the size of this jump r is of order

$$((t - s)|x|)^{1/(1+\beta)} \approx |x|^{3/(1+\beta)},$$

see Fig. 1. Therefore the convolution of the jump $r\delta_x$ with $p_{t-s}(x - \cdot) - p_{t-s}(0 - \cdot)$ is of order

$$|x|^{3/(1+\beta)}(p_{t-s}(0) - p_{t-s}(|x|)) \approx |x|^{3/(1+\beta)-1}.$$

In the last step we used (6.1). This leads then to the result that difference of values of the density at points x and 0 is of the same order. Then the pointwise Hölder exponent at 0 should be

$$\frac{3}{1+\beta} - 1 = \bar{\eta}_c.$$

This heuristic works for $\bar{\eta}_c < 1$. In the case $\bar{\eta}_c > 1$ the density becomes differentiable. For that reason one has to convolute $r\delta_x$ with $q_{t-s}(x - \cdot, 0 - \cdot)$, where q is defined in (2.22). This convolution is also of the order

$$|x|^{3/(1+\beta)}|q_{t-s}(0,x)| \approx |x|^{3/(1+\beta)-1}. \tag{6.2}$$

Again, we arrive at the same value of the pointwise Hölder exponent $\bar{\eta}_c$.

© The Author(s) 2016
L. Mytnik, V. Wachtel, *Regularity and Irregularity of Superprocesses with (1 + β)-stable Branching Mechanism*, SpringerBriefs in Probability and Mathematical Statistics, DOI 10.1007/978-3-319-50085-0_6

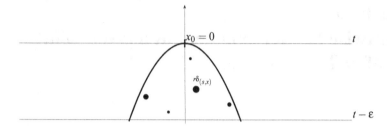

Fig. 1: The maximal jump that occurs at time s at a spatial position x with $|x| < (t-s)^{1/2}$ is of order $r \approx ((t-s)|x|)^{1/(1+\beta)}$. A sequence of such jumps at times $s_i \uparrow t$ determines the Hölder index of continuity of $X_t(\cdot)$ at the fixed spatial point $x_0 = 0$.

Since the case $\bar{\eta}_c < 1$ has been studied in [19], we shall concentrate here on the case $\bar{\eta}_c > 1$. In other words, we shall assume that $\beta < 1/2$. Under this assumption, the function $\frac{\partial}{\partial y} p_{t-u}(x_2 - y)$ is integrable with respect to $M(du, dy)$. Consequently, we may consider $Z_t(x_1, x_2)$ defined in (2.21) with q defined in (2.22).

6.1 Proof of the lower bound for $H_X(0)$

To get a lower bound for $H_Z(0)$ it suffices to show that, for every positive γ,

$$\sup_{0 < x < 1} \frac{|Z_t(x, 0)|}{|x|^{\bar{\eta}_c - \gamma}} < \infty$$

with probability one.

Let $\Delta Z_s(x_1, x_2)$ denote the jump of $Z_s(x_1, x_2)$:

$$\Delta Z_s(x_1, x_2) := Z_s(x_1, x_2) - Z_{s-}(x_1, x_2).$$

Denote

$$A_1^\varepsilon := \tag{6.3}$$
$$\left\{ \Delta X_s(y) \le c_{(4.14)} \left((t-s)|y| \right)^{1/(1+\beta)} (f_{s,x})^\ell \text{ for all } s < t \text{ and } y \in B_2(0) \right\}$$
$$\cap \left\{ \Delta X_s(y) \le c_{(4.5)} (t-s)^{1/(1+\beta)-\gamma} \text{ for all } s < t \text{ and } y \in \mathbb{R} \right\} \cap \{ V(B_2(0)) \le c_\varepsilon \},$$

with V defined in Lemma 3.5. According to Lemma 3.5, there exists c_ε such that

$$\mathbf{P}(V(B_2(0)) > c_\varepsilon) < \varepsilon. \tag{6.4}$$

Combining this with Lemma 4.2 and Lemma 4.3, we conclude that

$$\mathbf{P}(A_1^\varepsilon) > 1 - 3\varepsilon. \tag{6.5}$$

In the next lemma we derive an upper bound on jumps of $Z_s(x,0)$ on the set A_1^ε. This bound follows from the first and the second components in the definition of this set, which give bounds on jumps of X before time t. In order to transfer the bound on jumps of $Z_s(x,0)$ into a bound on values of $Z_s(x,0)$ we are going to use the representation of $Z_s(x,0)$ via time-changed $(1+\beta)$-stable processes, see Lemma 2.2. To this end we need a good control on the corresponding time change. This control is given by the component $\{V(B_2(0)) \le c_\varepsilon\}$ in the definition of A_1^ε. The resulting bound on the values of $Z_s(x,0)$ is derived in Lemma 6.2, which confirms the upper bound part of (6.2).

Lemma 6.1. *On the event A_1^ε we have, for all $s \le t$ and all $x \in \mathbb{R}$,*

$$|\Delta Z_s(x,0)| \le C|x|^{\overline{\eta}_c - 3\gamma}.$$

Proof. Let (y,s,r) be the point of an arbitrary jump of the measure $\mathcal{N}$ with $s \le t$. Then for the corresponding jump of $Z_s(x,0)$ we have the following bound:

$$|\Delta Z_s(x,0)| \le r|q_{t-s}(x-y,-y)|. \tag{6.6}$$

it follows from Corollary 4.4 that

$$r \le C(t-s)^{\frac{1}{1+\beta}-\gamma}|y|^{\frac{1}{1+\beta}-\gamma}$$

on the event A_1^ε.

Now we will use Lemma A.1 from Appendix A. Applying (A.4) from that lemma with $\delta = \overline{\eta}_c - 1 - 3\gamma$ to $|q_{t-s}(x-y,-y)|$, we obtain

$$|\Delta Z_s(x,0)|$$

$$\le C(t-s)^{\frac{1}{1+\beta}-\gamma}|y|^{\frac{1}{1+\beta}-\gamma}\frac{|x|^{\overline{\eta}_c-3\gamma}}{(t-s)^{(\overline{\eta}_c-3\gamma)/2}}\left(p_{t-s}(-y/2) + p_{t-s}((x-y)/2)\right)$$

$$= C|x|^{\overline{\eta}_c-3\gamma}\left(\frac{|y|}{(t-s)^{1/2}}\right)^{\frac{1}{1+\beta}-\gamma}\left(p_1\left(\frac{-y}{2(t-s)^{1/2}}\right) + p_1\left(\frac{x-y}{2(t-s)^{1/2}}\right)\right). \tag{6.7}$$

Assume first that $|x| \le (t-s)^{1/2}$. If $|y| > 2(t-s)^{1/2}$, then $|x-y| > |y|/2$ and, consequently,

$$\left(\frac{|y|}{(t-s)^{1/2}}\right)^{\frac{1}{1+\beta}-\gamma}\left(p_1\left(\frac{-y}{2(t-s)^{1/2}}\right) + p_1\left(\frac{x-y}{2(t-s)^{1/2}}\right)\right)$$

$$\le 2\left(\frac{|y|}{(t-s)^{1/2}}\right)^{\frac{1}{1+\beta}-\gamma}p_1\left(\frac{|y|}{4(t-s)^{1/2}}\right).$$

Noting that the function on the right-hand side is bounded, we conclude from (6.7) that

$$|\Delta Z_s(x,0)| \leq C|x|^{\overline{\eta}_c - 3\gamma} \tag{6.8}$$

for $|x| \leq (t-s)^{1/2}$ and $|y| > 2(t-s)^{1/2}$. Moreover, if $|y| \leq 2(t-s)^{1/2}$, then

$$\left(\frac{|y|}{(t-s)^{1/2}}\right)^{\frac{1}{1+\beta}-\gamma}\left(p_1\left(\frac{-y}{2(t-s)^{1/2}}\right)+p_1\left(\frac{x-y}{2(t-s)^{1/2}}\right)\right) \leq 2^{1+\beta}.$$

Consequently, (6.8) is valid also for $|y| \leq 2(t-s)^{1/2}$ and $|x| \leq (t-s)^{1/2}$.

Assume now that $|x| > (t-s)^{1/2}$ and $|y| > 2(t-s)^{1/2}$. In this case we bound q_{t-s} in a completely different way:

$$|q_{t-s}(x-y,-y)| \leq |p_{t-s}(x-y) - p_{t-s}(-y)| + |x|\left|\frac{\partial}{\partial y}p_{t-s}(-y)\right|.$$

Let us use again results from Appendix A. Applying (A.1) with $\delta = \eta_c - 2\gamma$ and (A.2), we see that $|q_{t-s}(x-y,-y)|$ is bounded by

$$C|x|^{\eta_c-2\gamma}(t-s)^{-\frac{1}{1+\beta}+\gamma}\left(p_1\left(\frac{-y}{2(t-s)^{1/2}}\right)+p_1\left(\frac{x-y}{2(t-s)^{1/2}}\right)\right)$$
$$+C\frac{|x|}{t-s}p_1\left(\frac{-y}{2(t-s)^{1/2}}\right).$$

Consequently,

$$|\Delta Z_s(x,0)| \leq C|y|^{\frac{1}{1+\beta}-\gamma}|x|^{\eta_c-2\gamma}\left(p_1\left(\frac{-y}{2(t-s)^{1/2}}\right)+p_1\left(\frac{x-y}{2(t-s)^{1/2}}\right)\right)$$
$$+C|x||y|^{\frac{1}{1+\beta}-\gamma}(t-s)^{\frac{1}{1+\beta}-1-\gamma}p_1\left(\frac{-y}{2(t-s)^{1/2}}\right). \tag{6.9}$$

Since $u^{\frac{1}{1+\beta}-\gamma}p_1(u)$ is bounded, the term in the second line in (6.9) does not exceed

$$C|x|(t-s)^{\frac{3}{2(1+\beta)}-1-\frac{3\gamma}{2}}.$$

As a result, for $|x| > (t-s)^{1/2}$ we get

$$|y|^{\frac{1}{1+\beta}-\gamma}(t-s)^{\frac{1}{1+\beta}-1-\gamma}p_1\left(\frac{-y}{2(t-s)^{1/2}}\right) \leq C|x|^{\frac{3}{1+\beta}-1-3\gamma} = C|x|^{\overline{\eta}_c-3\gamma}. \tag{6.10}$$

By the same argument,

$$|y|^{\frac{1}{1+\beta}-\gamma}|x|^{\eta_c-2\gamma}p_1\left(\frac{-y}{2(t-s)^{1/2}}\right) \leq C|x|^{\overline{\eta}_c-3\gamma}. \tag{6.11}$$

We next note that if $|y| > 2|x|$, then $|y - x| > |y|/2$ and, consequently,

$$p_1\left(\frac{x-y}{2(t-s)^{1/2}}\right) \leq p_1\left(\frac{-y}{4(t-s)^{1/2}}\right).$$

Therefore,

$$|y|^{\frac{1}{1+\beta}-\gamma}|x|^{\eta_c-2\gamma}p_1\left(\frac{x-y}{2(t-s)^{1/2}}\right) \leq C|x|^{\overline{\eta}_c-3\gamma} \tag{6.12}$$

for $|y| > 2|x|$. But if $|y| \leq 2|x|$, then (6.12) is obvious. Combining (6.9)–(6.12) we conclude that (6.8) holds for $|x| > (t-s)^{1/2}$. This completes the proof. $\qquad\square$

Lemma 6.2. *For every fixed x with $|x| < 1$ we have*

$$\mathbf{P}\left(|Z_t(x,0)| > r|x|^{\overline{\eta}_c-2\gamma}, A_1^\varepsilon\right) \leq \left(\frac{C}{r}|x|^\gamma\right)^{r|x|^{-\gamma}/C}, \qquad r > 0.$$

Proof. According to Lemma 2.2 there exist spectrally positive $(1+\beta)$-stable processes $L^+(x)$ and $L^-(x)$ such that

$$Z_t(x,0) = L^+_{T_+(x)} - L^-_{T_-(x)}, \tag{6.13}$$

where

$$T_\pm(x) := \int_0^t du \int_{\mathbb{R}} X_u(dy)\left((q_{t-u}(x-y,-y))^\pm\right)^{1+\beta}.$$

By (3.5) with $\theta = 1+\beta$ and $\delta = (1-2\beta-\gamma)/(1+\beta)$, there exists $C = C(\varepsilon,\gamma)$ such that

$$T_\pm(x) \leq C|x|^{2-\beta-\varepsilon_1} \quad \text{on } A_1^\varepsilon. \tag{6.14}$$

Applying Lemma 6.1 and (6.14), we obtain

$$\mathbf{P}\left(L^\pm_{T_\pm(x)}(x) \geq r|x|^{\overline{\eta}_c-2\gamma}, A_1^\varepsilon\right)$$

$$\leq \mathbf{P}\left(L^\pm_{T_\pm(x)}(x) \geq r|x|^{\overline{\eta}_c-2\gamma}, A_1^\varepsilon, \sup_{s\leq T_\pm(x)}\Delta L_s^\pm \leq C|x|^{\overline{\eta}_c-3\gamma}\right)$$

$$\leq \mathbf{P}\left(\sup_{s\leq C|x|^{2-\beta-\varepsilon_1}} L_s^\pm(x)1\{\sup_{0\leq v\leq s}\Delta L_v^\pm \leq C|x|^{\overline{\eta}_c-3\gamma}\} \geq r|x|^{\overline{\eta}_c-2\gamma}\right).$$

In Appendix B we derive upper bounds for tails of the distribution of a spectrally positive κ-stable process. In particular, in Lemma B.1 we prove a probability inequality for the process in the absence of big jumps. Using this lemma with $\kappa = 1+\beta$, we get

$$\mathbf{P}\left(L^{\pm}_{T_{\pm}(x)}(x) \geq r|x|^{\overline{\eta}_c - 2\gamma}, A_1^{\varepsilon}\right) \leq \left(\frac{C|x|^{2-\beta-\gamma}}{r|x|^{\overline{\eta}_c - 2\gamma}|y|^{\beta(\overline{\eta}_c - 3\gamma)}}\right)^{r|x|^{-\gamma}/C}$$

$$= \left(\frac{C}{r}|x|^{\gamma}\right)^{r|x|^{-\gamma}/C}.$$

Thus, the proof is finished.   □

Taking into account (6.5), we see that the lower bound for $H_Z(0)$ will be proven if we show that, for every $\varepsilon > 0$,

$$\sup_{0 < x < 1} \frac{|Z_t(x,0)|}{|x|^{\overline{\eta}_c - \gamma}} < \infty \quad \text{on } A_1^{\varepsilon}. \tag{6.15}$$

Fix some $q \in (0,1)$ and note that

$$\left\{\sup_{0 < x < 1} \frac{|Z_t(x,0)|}{x^{\overline{\eta}_c - \gamma}} > k\right\} \subseteq \bigcup_{n=1}^{\infty} \left\{\sup_{x \in I_n} |Z_t(x,0)| > \frac{k}{2^q} n^{-q(\overline{\eta}_c - \gamma)}\right\},$$

where $I_n := \{x : (n+1)^{-q} \leq x < n^{-q}\}$. Moreover, by the triangle inequality,

$$|Z_t(x,0)| \leq |Z_t(n^{-q},0)| + |Z_t(x) - Z_t(n^{-q})| + (n^{-q} - x)|W|, \quad x \in I_n,$$

where

$$W := \int_0^t \int_{\mathbb{R}} M(\mathrm{d}(u,y)) \frac{\partial}{\partial y} p_{t-u}(-y).$$

Consequently, for all $n \geq 1$,

$$\left\{\sup_{x \in I_n} |Z_t(x,0)| > \frac{k}{2^q} n^{-q(\bar{\eta}_c - \gamma)}\right\} \subseteq \left\{\sup_{x \in I_n} |Z_t(x) - Z_t(n^{-q})| > \frac{k}{3 \cdot 2^q} n^{-q}\right\}$$

$$\cup \left\{|Z_t(n^{-q},0)| > \frac{k}{3 \cdot 2^q} n^{-q(\bar{\eta}_c - \gamma)}\right\} \cup \left\{|W| > \frac{kn^{-q(\bar{\eta}_c - \gamma)}}{3 \cdot 2^q (n^{-q} - (n+1)^{-q})}\right\},$$

Note that, for all $R > 0$,

$$\left\{\sup_{0 < x < y < 1} \frac{|Z_t(x) - Z_t(y)|}{|x-y|^{q(\bar{\eta}_c - \gamma)/(q+1)}} \leq R\right\}$$

$$\subseteq \left\{|Z_t(x) - Z_t(n^{-q})| \leq Rq^{q(\bar{\eta}_c - \gamma)/(q+1)} n^{-q\eta}, \ x \in I_n\right\}.$$

This implies that

$$\left\{ \sup_{0<x<1} \frac{|Z_t(x,0)|}{x^{\bar{\eta}_c-\gamma}} > k \right\} \subseteq \left\{ \sup_{0<x<y<1} \frac{|Z_t(x) - Z_t(y)|}{|x-y|^{q(\bar{\eta}_c-\gamma)/(q+1)}} > c(q)k \right\}$$

$$\cup \bigcup_{n=1}^{\infty} \left\{ |Z_t(n^{-q},0)| > \frac{k}{3\cdot 2^q} n^{-q(\bar{\eta}_c-\gamma)} \right\} \cup \{|W| > c(q)k\}.$$

where $c(q)$ is some positive constant.

If we choose q so small that $(\bar{\eta}_c - \gamma)q/(q+1) < \eta_c$, then

$$\lim_{k\to\infty} \mathbf{P}\left(\sup_{0<x<y<1} \frac{|Z_t(x) - Z_t(y)|}{|x-y|^{q(\bar{\eta}_c-\gamma)/(q+1)}} > c(q)k \right) = 0,$$

since, by Theorem 1.2, Z is locally Hölder continuous of every index smaller than η_c. Furthermore,

$$\lim_{k\to\infty} \mathbf{P}(|W| > c(q)k) = 0.$$

Finally, applying Lemma 6.2, conclude that

$$\lim_{k\to\infty} \mathbf{P}\left(\bigcup_{n=1}^{\infty} \left\{ |Z_t(n^{-q},0)| > \frac{k}{3\cdot 2^q} n^{-q(\bar{\eta}_c-\gamma)} \right\} \right) = 0.$$

Thus, (6.15) is shown.

6.2 Proof of the optimality of $\bar{\eta}_c$

Now it is time to explain the detailed strategy of the optimality proof. First note that it follows from the definition (4.21) of the set D_θ that $D_\theta \uparrow \{X_t(0) > 0\}$ as $\theta \downarrow 0$. This and (6.5) imply that for the proof of Theorem 1.3(b) *it is sufficient* to show that

$$\mathbf{P}\left(\sup_{x\in B_\varepsilon(0), x\neq 0} \frac{|X_t(x) - X_t(0)|}{|x|^{\bar{\eta}_c}} = \infty \middle| D_\theta \cap A_1^\varepsilon \right) = 1$$

for all sufficiently small $\varepsilon > 0$.

Since $\mu * p_t(x)$ is Lipschitz continuous at 0, the latter will follow from the equality

$$\mathbf{P}\left(Z_t(c2^{-n-2},0) \geq 2^{-\bar{\eta}_c n} n^{1/(1+\beta)-\varepsilon} \text{ infinitely often} \middle| D_\theta \cap A_1^\varepsilon \right) = 1, \quad (6.16)$$

where we choose $c = c_{(4.22)}$.

To verify (6.16), we will again use the method of representing Z as a time-changed stable process. To be more precise, applying (6.13) with $x = c2^{-n-2}$ (for n sufficiently large) and using n-dependent notation as $L_n^\pm, T_{n,\pm}$ (and $\varphi_{n,\pm}$), we have

$$Z_t(c2^{-n-2}, 0) = L_n^+(T_{n,+}) - L_n^-(T_{n,-}).$$

It follows from this representation that (6.16) is a consequence in the following statement:

Proposition 6.3. *For almost every* $\omega \in D_\theta \cap A_1^\varepsilon$ *there exists a subsequence* n_j *such that*

$$L_{n_j}^+(T_{n_j,+}) \geq 2^{1-\bar{n}_c n_j} n_j^{1/(1+\beta)-\varepsilon} \quad and \quad L_{n_j}^-(T_{n_j,-}) \leq 2^{-\bar{n}_c n_j} n_j^{1/(1+\beta)-\varepsilon}.$$

Let us define the following events:

$$B_n^+ := \left\{ L_n^+(T_{n,+}) \geq 2^{1-\bar{n}_c n} n^{1/(1+\beta)-\varepsilon} \right\}, \quad B_n^- := \left\{ L_n^-(T_{n,-}) \leq 2^{-\bar{n}_c n} n^{1/(1+\beta)-\varepsilon} \right\}$$

and

$$B_n := B_n^+ \cap B_n^-.$$

Then, obviously, Proposition 6.3 will follow once we verify

$$\lim_{N\uparrow\infty} \mathbf{P}\left(\bigcup_{n=N}^{\infty} (B_n \cap A_n) \,\Big|\, D_\theta \cap A_1^\varepsilon \right) = 1, \tag{6.17}$$

where A_n were defined in (4.26). Taking into account Lemma 4.6, we conclude that to get (6.17) we have to show

$$\lim_{N\uparrow\infty} \mathbf{P}\left(\bigcup_{n=N}^{\infty} (B_n^c \cap A_n) \,\Big|\, D_\theta \cap A_1^\varepsilon \right) = 0. \tag{6.18}$$

Let us explain briefly the meaning of (6.17). By Lemma 4.6 we know that there exists a sequence of big jumps

$$\Delta X_s\left(\frac{c}{2} 2^{-n_j}, \frac{3c}{2} 2^{-n_j} \right) \geq 2^{-(\bar{n}_c+1)n_j} n_j^{1/(1+\beta)}$$

for some $s \in (t - 2^{-2n_j}, t - 2^{-2(n_j+1)})$. (6.17) implies that these jumps guarantee big values of $L_{n_j}^+(T_{n_j,+}) - L_{n_j}^-(T_{n_j,-})$ for some subsequence of $\{n_j\}$. And this is the main consequence of Proposition 6.3.

Now we will present two lemmas, from which (6.18) will follow immediately. To this end, split

$$B_n^c \cap A_n = (B_n^{+,c} \cap A_n) \cup (B_n^{-,c} \cap A_n). \tag{6.19}$$

Lemma 6.4. *We have*

$$\lim_{N\uparrow\infty} \sum_{n=N}^{\infty} \mathbf{P}\big(B_n^{+,c} \cap A_n \cap A_1^\varepsilon\big) = 0.$$

The proof of this lemma is a word-for-word repetition of the proof of Lemma 5.3 in [18] (it is even simpler as we do not need additional indexing in k here), and we omit it. The idea behind the proof is simple: Whenever X has a "big" jump guaranteed by A_n, this jump corresponds to the jump of L_n^+ and then it is very difficult for a spectrally positive process L_n^+ to come down, which is required by $B_n^{+,c}$.

Note that Lemma 6.4 alone is not enough to finish the proof of Proposition 6.3: on $\{L_n^+ \geq 2^{1-\bar{\eta}_c n} n^{1/(1+\beta)-\varepsilon}\}$ we may still have $Z_t(c2^{-n-2},0) \leq 2^{-\bar{\eta}_c n} n^{1/(1+\beta)-\varepsilon}$ if $B_n^{-,c}$ occurs. In other words, this is the situation where the jump of X, which leads to a large value of L_n^+, can be compensated by a further jump of X. The next lemma states that it cannot happen that all the jumps guaranteed by A_n's will be compensated.

Lemma 6.5. *We have*

$$\lim_{N\uparrow\infty} \sum_{n=N}^{\infty} \mathbf{P}\big(B_n^{-,c} \cap A_n \cap A_1^\varepsilon \cap D_\theta\big) = 0.$$

Now we are ready to finish

Proof of Proposition 6.3. Combining Lemmata 6.4 and 6.5, we conclude that there exists a subsequence $\{n_j\}$ with properties described in Proposition 6.3. $\qquad\square$

The remaining part of the book will be devoted to the proof of Lemma 6.5 and we prepare now for it.

One can easily see that $B_n^{-,c}$ is a subset of a union of two events (with the obvious correspondence):

$$B_n^{-,c} \subseteq U_n^1 \cup U_n^2 := \big\{\Delta L_n^- > 2^{-\bar{\eta}_c n} n^{1/(1+\beta)-2\varepsilon}\big\}$$
$$\cup \big\{\Delta L_n^- \leq 2^{-\bar{\eta}_c n} n^{1/(1+\beta)-2\varepsilon}, \ L_n^-(T_{n,-}) > 2^{-\bar{\eta}_c n} n^{1/(1+\beta)-\varepsilon}\big\},$$

where

$$\Delta L_n^- := \sup_{0<s\leq T_{n,-}} \Delta L_n^-(s).$$

The occurrence of the event U_n^1 means that L_n^- has big jumps. If U_n^2 occurs, it means that L_n^- gets large without big jumps. According to Appendix B, stable processes without big jumps can achieve large values only with quite small probability. Thus, the statement of the next lemma is not surprising.

Lemma 6.6. *We have*

$$\lim_{N\uparrow\infty} \sum_{n=N}^{\infty} \mathbf{P}(U_n^2 \cap A_1\varepsilon) = 0.$$

We omit the proof of this lemma as well, since its crucial part related to bounding of $\mathbf{P}(U_n^2 \cap A_1^\varepsilon)$ is a repetition of the proof of Lemma 5.6 in [18] (again with obvious simplifications).

Lemma 6.7. *There exist constants ρ and ξ such that, for all sufficiently large values of n,*

$$A_1^\varepsilon \cap A_n \cap U_n^1 \subseteq A_1^\varepsilon \cap E_n(\rho, \xi),$$

where

$$E_n(\rho, \xi) := \left\{ \text{There exist at least two jumps of } M \text{ of the form } r\delta_{(s,y)} \text{ such that} \right.$$

$$r \geq \left((t-s) \max\{(t-s)^{1/2}, |y|\}\right)^{1/(1+\beta)} \log^{1/(1+\beta)-2\varepsilon}\left((t-s)^{-1}\right),$$

$$\left. |y| \leq (t-s)^{1/2} \log^\xi\left((t-s)^{-1}\right), \quad s \in \left[t - 2^{-2n} n^\rho, t - 2^{-2n} n^{-\rho}\right] \right\}.$$

Proof. By the definition of A_n, there exists a jump of M of the form $r\delta_{(s,y)}$ with r, s as in $E_n(\rho, \xi)$, and $y > c2^{-n-1}$. Furthermore, noting that $\varphi_{n,-}(y) = 0$ for $y \geq c2^{-n-3}$, we see that the jumps $r\delta_{(s,y)}$ of M contribute to $L_n^-(T_{n,-})$ if and only if $y < c2^{-n-3}$. Thus, to prove the lemma it is sufficient to show that U_n^1 yields the existence of at least one further jump of M on the half-line $\{y < c2^{-n-3}\}$ with properties mentioned in the statement. Denote

$$D := \left\{ (r,s,y) : r \geq \left((t-s) \max\{(t-s)^{1/2}, |y|\}\right)^{1/(1+\beta)} \log^{1/(1+\beta)-2\varepsilon}\left((t-s)^{-1}\right), \right.$$

$$y \in \left(-(t-s)^{1/2} \log^\xi\left((t-s)^{-1}\right), c2^{-n-3}\right),$$

$$\left. s \in \left[t - 2^{-2n} n^\rho, t - 2^{-2n} n^{-\rho}\right] \right\}.$$

$$(6.20)$$

Then we need to show that U_n^1 implies the existence of a jump $r\delta_{(s,y)}$ of M with $(r,s,y) \in D$.

Note that

$$D = D_1 \cap D_2 \cap D_3$$

$$:= \left\{ (r,s,y) : r \geq 0, \ s \in [0,t], \ y \in \left(-(t-s)^{1/2} \log^\xi\left((t-s)^{-1}\right), c2^{-n-3}\right) \right\}$$

$$\cap \left\{ (r,s,y) : r \geq 0, \ y \in (-\infty, c2^{-n-3}), \ s \in \left[t - 2^{-2n} n^\rho, t - 2^{-2n} n^{-\rho}\right] \right\}$$

$$\cap \left\{ (r,s,y) : y \in (-\infty, c2^{-n-3}), \ s \in [0,t], \right.$$

$$\left. r \geq \left((t-s) \max\{(t-s)^{1/2}, |y|\}\right)^{1/(1+\beta)} \log^{1/(1+\beta)-2\varepsilon}\left((t-s)^{-1}\right) \right\}.$$

Therefore,

$$D^c \cap \{y < c2^{-n-3}\} = \left(D_1^c \cap \{y < c2^{-n-3}\}\right) \cup (D_1 \cap D_2^c) \cup (D_1 \cap D_2 \cap D_3^c),$$

where the complements are defined with respect to the set

$$\{(r,s,y) : r \geq 0, \ s \in [0,t], \ y \in \mathbb{R}\}.$$

We first show that any jumps of M in $D_1^c \cap \{y < c2^{-n-3}\}$ cannot be the course of a jump of L_n^- such that U_n^1 holds. Indeed, using the last inequality in Lemma A.1 from Appendix A with $\delta = \bar{\eta}_c$, we get for $y < c2^{-n-3}$ the inequality

$$\begin{aligned}
(q_{t-s}(c2^{-n-2}, 0))^- &\leq C2^{-\bar{\eta}_c n}(t-s)^{-\bar{\eta}_c/2} p_{t-s}(y/2) \\
&\leq C2^{-\bar{\eta}_c n}(t-s)^{-(1+\bar{\eta}_c)/2} \exp\left\{-\frac{y^2}{8(t-s)}\right\} \\
&\leq C2^{-\bar{\eta}_c n}(t-s)^{1-\bar{\eta}_c/2} |y|^{-3},
\end{aligned} \tag{6.21}$$

in the second step we used the scaling property of the kernel p, and in the last step we have used the trivial bound $e^{-x} \leq x^{-3/2}$.

Further, by (6.3), on the set A_1^ε we have

$$\Delta X_s(y) \leq C\left(|y|(t-s)\right)^{1/(1+\beta)} (f_{s,y})^\ell, \quad |y| \leq 1/e, \tag{6.22}$$

and

$$\Delta X_s(y) \leq C(t-s)^{1/(1+\beta)-\gamma}, \quad |y| > 1/e, \tag{6.23}$$

and recall that $f_{s,x} = \log\left((t-s)^{-1}\right) 1_{\{x \neq 0\}} \log(|x|^{-1})$. Combining (6.21) and (6.22), we conclude that the corresponding jump of L_n^-, henceforth denoted by $\Delta L_n^-[r\delta_{(s,y)}]$, is bounded by

$$C2^{-\bar{\eta}_c n}(t-s)^{1-\bar{\eta}_c/2+\frac{1}{1+\beta}} \log^{\frac{1}{1+\beta}+q}\left((t-s)^{-1}\right) |y|^{-3+\frac{1}{1+\beta}} \log^{\frac{1}{1+\beta}+q}\left(|y|^{-1}\right).$$

Since $|y|^{-3+\frac{1}{1+\beta}} \log^{\frac{1+\gamma}{1+\beta}}\left(|y|^{-1}\right)$ is monotone decreasing, we get, maximizing over y, for $y < -(t-s)^{1/2} \log^\xi\left((t-s)^{-1}\right)$ the bound

$$\Delta L_n^-[r\delta_{(s,y)}] \leq C2^{-\bar{\eta}_c n} \log^{\frac{2}{1+\beta}+2q-\xi(3-\frac{1}{1+\beta})}\left(|y|^{-1}\right).$$

Choosing $\xi \geq \frac{2+2q(1+\beta)}{3(1+\beta)-1}$, we see that

$$\Delta L_n^-[r\delta_{(s,y)}] \leq C2^{-\bar{\eta}_c n}, \quad |y| < 1/e. \tag{6.24}$$

Moreover, if $y < -1/e$, then it follows from (6.21) and (6.23) that the jump $\Delta L_n^-[r\delta_{(s,y)}]$ is bounded by

$$C2^{-\bar{\eta}_c n}(t-s)^{1-\bar{\eta}_c/2+\frac{1}{1+\beta}-\gamma} |y|^{-3} \leq C2^{-\bar{\eta}_c n}. \tag{6.25}$$

Combining (6.24) and (6.25), we see that all the jumps of M in $D_1^c \cap \{y < c2^{-n-3}\}$ do not produce jumps of L_n^- such that U_n^1 holds.

We next assume that M has a jump $r\delta_{(s,y)}$ in $D_1 \cap D_2^c$. If, additionally, $s \le t - 2^{-2n} n^{\rho}$, then, using the last inequality in Lemma A.1 (from Appendix A) with $\delta = 1$, we get

$$(q_{t-s}(c2^{-n-2},0))^- \le C2^{-2n}(t-s)^{-3/2}.$$

From this bound and (6.22) we obtain

$$\Delta L_n^-[r\delta_{(s,y)}] \le C2^{-2n}(t-s)^{-3/2+\frac{1}{1+\beta}}\log^{\frac{1}{1+\beta}+q}((t-s)^{-1})|y|^{\frac{1}{1+\beta}}\log^{\frac{1}{1+\beta}+q}(|y|^{-1})$$
$$\le C2^{-2n}(t-s)^{\frac{3}{2}(\frac{1}{1+\beta}-1)}\log^{\frac{2+\xi}{1+\beta}+2q}((t-s)^{-1}).$$

Using the assumption $t - s \ge 2^{-2n} n^{\rho}$, we arrive at the inequality

$$\Delta L_n^-[r\delta_{(s,y)}] \le C2^{-\bar{\eta}_c n}n^{\frac{2+\xi}{1+\beta}+2q+\frac{\rho}{2}(\bar{\eta}_c-2)}.$$

From this we see that if we choose $\rho \ge \frac{2(2q+(2+\xi)/(1+\beta))}{2-\bar{\eta}_c}$, then the jumps of L_n^- are bounded by $C2^{-\bar{\eta}_c n}$, and hence U_n^1 does not occur.

Using (A.1) with $\delta = 1$ and (A.2) from Appendix A, one can easily derive

$$(q_{t-s}(c2^{-n-2},0))^- \le C2^{-n}(t-s)^{-1}.$$

Then for $y \in \left(-(t-s)^{1/2}\log^{\xi}((t-s)^{-1}), c2^{-n-3}\right)$ and $t - s \le 2^{-2n} n^{-\rho}$ we have the inequality

$$\Delta L_n^-[r\delta_{(s,y)}] \le C2^{-n}(t-s)^{-1}(|y|(t-s))^{\frac{1}{1+\beta}}(f_{s,y})^{\ell}$$
$$\le C2^{-n}(t-s)^{\frac{3}{2(1+\beta)}-1}\log^{\frac{2+\xi}{1+\beta}+2q}((t-s)^{-1})$$
$$= C2^{-n}(t-s)^{(\bar{\eta}_c-1)/2}\log^{\frac{2+\xi}{1+\beta}+2q}((t-s)^{-1})$$
$$\le C2^{-\bar{\eta}_c n}n^{-\rho(\bar{\eta}_c-1)/2+\frac{2+\xi}{1+\beta}+2q}.$$

Choosing $\rho \ge \frac{2(\xi+2+2q(1+\beta))}{(1+\beta)(\bar{\eta}_c-1)}$, we conclude that $\Delta L_n^-[r\delta_{(s,y)}] \le C2^{-\bar{\eta}_c n}$, and again U_n^1 does not occur.

Finally, it remains to consider the jumps of M in $D_1 \cap D_2 \cap D_3^c$. If the value of the jump does not exceed $(t-s)^{\frac{3}{2(1+\beta)}}\log^{\frac{1}{1+\beta}-2\varepsilon}((t-s)^{-1})$, then it follows from (A.4) (in Appendix A) with $\delta = \bar{\eta}_c - 1$ that

$$\Delta L_n^-[r\delta_{(s,y)}] \le C2^{-\bar{\eta}_c n}(t-s)^{-(\bar{\eta}_c-1)/2}(t-s)^{\frac{3}{2(1+\beta)}}\log^{\frac{1}{1+\beta}-2\varepsilon}((t-s)^{-1})$$
$$\le C2^{-\bar{\eta}_c n}\log^{\frac{1}{1+\beta}-2\varepsilon}((t-s)^{-1}).$$

Then, on D_2, that is, for $t - s > 2^{-2n} n^{-\rho}$,

$$\Delta L_n^- [r\delta_{(s,y)}] \le C 2^{-\bar{n}_c n} n^{\frac{1}{1+\beta} - 2\varepsilon}. \tag{6.26}$$

Furthermore, if $y < -(t-s)^{1/2}$ and the value of the jump is less than $\left(|y|(t-s)\right)^{\frac{1}{1+\beta}}$ $\log^{\frac{1}{1+\beta} - 2\varepsilon}\left((t-s)^{-1}\right)$, then, using (6.21), we get

$$\Delta L_n^- [r\delta_{(s,y)}] \le C 2^{-\bar{n}_c n} (t-s)^{1-\bar{n}_c/2} \log^{\frac{1}{1+\beta} - 2\varepsilon}\left((t-s)^{-1}\right) |y|^{-3 + \frac{1}{1+\beta}}$$
$$\le C 2^{-\bar{n}_c n} \log^{\frac{1}{1+\beta} - 2\varepsilon}\left((t-s)^{-1}\right).$$

Then, on D_2, that is, for $t - s > 2^{-2n} n^{-\rho}$,

$$\Delta L_n^- [r\delta_{(s,y)}] \le C 2^{-\bar{n}_c n} n^{\frac{1}{1+\beta} - 2\varepsilon}. \tag{6.27}$$

By (6.26) and (6.27), we see that the jumps of M in $D_1 \cap D_2 \cap D_3^c$ do not produce jumps such that U_n^1 holds. Combining all the above we conclude that to have $\Delta L_n^- [r\delta_{(s,y)}] > C 2^{-\bar{n}_c n} n^{\frac{1}{1+\beta} - 2\varepsilon}$ it is necessary to have a jump in $D_1 \cap D_2 \cap D_3$. Thus, the proof is finished.   □

Now we are ready to finish

Proof of Lemma 6.5. In view of the Lemmas 6.6 and 6.7, it suffices to show that

$$\lim_{N \uparrow \infty} \sum_{n=N}^{\infty} \mathbf{P}\left(E_n(\rho, \xi) \cap A_1^\varepsilon \cap D_\theta\right) = 0. \tag{6.28}$$

The intensity of the jumps in D [the set defined in (6.20) and satisfying conditions in $E_n(\rho, \xi)$] is given by

$$\int_{t-2^{-2n}n\rho}^{t-2^{-2n}n^{-\rho}} ds \int_{|y| \le (t-s)^{1/2} \log^\xi \left((t-s)^{-1}\right)} X_s(dy) \tag{6.29}$$
$$\frac{\log^{2\varepsilon(1+\beta)-1}\left((t-s)^{-1}\right)}{(t-s) \max\left\{(t-s)^{1/2}, |y|\right\}}.$$

Since in (6.28) we are interested in a limit as $N \uparrow \infty$, we may assume that n is such that $(t-s)^{1/2} \log^\xi \left((t-s)^{-1}\right) \le 1$ for $s \ge t - 2^{-2n} n^\rho$. We next note that

$$\int_{|y| \le (t-s)^{1/2}} \frac{X_s(dy)}{\max\left\{(t-s)^{1/2}, |y|\right\}}$$
$$= (t-s)^{-1/2} X_s\left((-(t-s)^{1/2}, (t-s)^{1/2})\right) \le \theta^{-1}$$

on D_θ. Further, for every $j \geq 1$ satisfying $j \leq \log^\xi \left((t-s)^{-1} \right)$,

$$\int_{j(t-s)^{1/2} \leq |y| \leq (j+1)(t-s)^{1/2}} \frac{X_s(dy)}{\max\{(t-s)^{1/2}, |y|\}}$$
$$\leq j^{-1}(t-s)^{-1/2} X_s \left(\{ y : \ j(t-s)^{1/2} \leq |y| \leq (j+1)(t-s)^{1/2} \} \right).$$

Since the set $\{ y : \ j(t-s)^{1/2} \leq |y| \leq (j+1)(t-s)^{1/2} \}$ is the union of two balls with radius $\frac{1}{2}(t-s)^{-1/2}$ and centers in $B_2(0)$, we can apply Lemma 3.6 with $c = 1$ to get

$$\int_{j(t-s)^{1/2} \leq |y| \leq (j+1)(t-s)^{1/2}} \frac{X_s(dy)}{\max\{(t-s)^{1/2}, |y|\}} \leq 2\theta^{-1} j^{-1}$$

on D_θ. As a result, on the event D_θ we get the inequality

$$\int_{|y| \leq (t-s)^{1/2} \log^\xi \left((t-s)^{-1} \right)} X_s(dy) \frac{1}{\max\{(t-s)^{1/2}, |y|\}}$$
$$\leq C\theta^{-1} \log \left(\left| \log \left((t-s)^{-1} \right) \right| \right).$$

Substituting this into (6.29), we conclude that the intensity of the jumps is bounded by

$$C\theta^{-1} \int_{t-2^{-2n}n^\rho}^{t-2^{-2n}n^{-\rho}} ds \ \frac{\log^{2\varepsilon(1+\beta)-1} \left((t-s)^{-1} \log\log \left((t-s)^{-1} \right) \right)}{(t-s)}.$$

Simple calculations show that the latter expression is less than

$$C\theta^{-1} n^{2\varepsilon(1+\beta)-1} \log^{1+2\varepsilon(1+\beta)} n.$$

Consequently, since $E_n(\rho, \xi)$ holds when there are two jumps in D, we have

$$\mathbf{P}\left(E_n(\rho, \xi) \cap A_1^\varepsilon \cap D_\theta \right) \leq C\theta^{-2} n^{4\varepsilon(1+\beta)-2} \log^{2+4\varepsilon(1+\beta)} n.$$

Because $\varepsilon < 1/8 \leq 1/4(1+\beta)$, the sequence $\mathbf{P}\left(E_n(\rho, \xi) \cap A_1^\varepsilon \cap D_\theta \right)$ is summable, and the proof of the lemma is complete. □

Chapter 7
Elements of the proof of Theorem 1.5 and Proposition 1.6

The spectrum of singularities of X_t coincides with that of Z. Consequently, to prove Theorem 1.5, we have to determine Hausdorff dimensions of the sets

$$\mathscr{E}_{Z,\eta} := \{x \in (0,1) : H_Z(x) = \eta\},$$
$$\tilde{\mathscr{E}}_{Z,\eta} := \{x \in (0,1) : H_Z(x) \leq \eta\}$$

and this is done in the next two sections.

As usual, we give some heuristic arguments, which should explain the result (1.8). Using heuristic arguments that led to (6.2), one can easily show that a jump of order $(t-s)^\nu$ occurring between times s and t implies that (if there are no other "big" jumps nearby) the pointwise Hölder exponent should be $2\nu - 1$ in the ball of radius $(t-s)^{1/2}$ centered at the spatial position of this jump. In other words, in order to have $H_X(x) = \eta$ at a point x, a jump of order $(t-s)^{(\eta+1)/2}$ should appear, whose distance to x is less or equal to $(t-s)^{1/2}$.

From the formula for the compensator we infer that the number of such jumps, $\mathbf{N}_\eta$, is of order

$$\mathbf{N}_\eta \approx \int_{t-s}^t duX_u((0,1)) \int_{(t-s)^{(\eta+1)/2}} r^{-2-\beta} dr$$
$$\approx (t-s)^{-(\eta+1)(1+\beta)/2} \int_{t-s}^t duX_u((0,1)).$$

It turns out that the random measure X_u can be replaced by the Lebesgue measure multiplied by a random factor. (The proof of this fact is one of the main technical difficulties.) As a result, the number of jumps $\mathbf{N}_\eta$, leading to pointwise Hölder exponent η at certain points, is of order

$$(t-s)^{-(\eta+1)(1+\beta)/2+1},$$

© The Author(s) 2016

L. Mytnik, V. Wachtel, *Regularity and Irregularity of Superprocesses with (1 + β)-stable Branching Mechanism*, SpringerBriefs in Probability and Mathematical Statistics, DOI 10.1007/978-3-319-50085-0_7

see Fig. 1 for schematic description of these jumps. Since every such jump effects the regularity in the ball of radius $\mathbf{r} \approx (t-s)^{1/2}$, the corresponding Hausdorff dimension α of the set of points x, with $H_X(x) = \eta$, can be obtained from the relation

$$\mathbf{N}_\eta \mathbf{r}^\alpha \approx ((t-s)^{1/2})^\alpha (t-s)^{-(\eta+1)(1+\beta)/2+1} \approx 1.$$

This gives

$$\alpha = (\eta+1)(1+\beta) - 2 = (\eta - \eta_c)(1+\beta),$$

which coincides with (1.8).

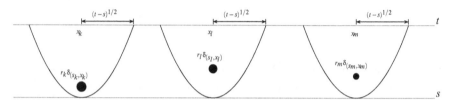

Fig. 1: If a jump occurring between times s and t has size r_i of order $(t-s)^{(\eta+1)/2}$ (and there are no "bigger" jumps nearby), then this jump determines the Hölder exponent η of $X_t(\cdot)$ in the ball of radius $(t-s)^{1/2}$ around the spatial position x_i of the jump. There are $\mathbf{N}_\eta \approx (t-s)^{-(\eta+1)(1+\beta)/2+1}$ such jumps in $[s,t] \times [0,1]$.

In the rest of this chapter we will justify the above heuristics.

7.1 Upper bound for the Hausdorff dimension: proof of Proposition 1.6

The aim of this section is to prove the following result, which trivially implies Proposition 1.6.

Proposition 7.1. *With probability one,*

$$\dim(\mathscr{E}_{Z,\eta}) \le \dim(\tilde{\mathscr{E}}_{Z,\eta}) \le (1+\beta)(\eta - \eta_c), \text{ for all } \eta \in [\eta_c, \overline{\eta}_c]. \tag{7.1}$$

We need to introduce an additional notation. In what follows, for any $\eta \in (\eta_c, \overline{\eta}_c) \setminus \{1\}$, we fix an arbitrary small $\gamma = \gamma(\eta) \in (0, \frac{10^{-2}\eta_c}{\alpha})$ such that

$$\gamma < \begin{cases} 10^{-2}\min\{1-\eta, \eta\}, & \text{if } \eta < 1, \\ 10^{-2}\min\{\eta-1, 2-\eta\}, & \text{if } \eta > 1, \end{cases}$$

and define

$$S_\eta := \Big\{ x \in (0,1) : \text{there exists a sequence } (s_n, y_n) \to (t, x)$$

$$\text{with } \Delta X_{s_n}(\{y_n\}) \ge (t-s_n)^{\frac{1}{1+\beta} - \gamma}|x - y_n|^{\eta - \eta_c} \Big\}.$$

To prove the above proposition we have to verify the following two lemmas:

Lemma 7.2. *For every* $\eta \in (\eta_c, \overline{\eta}_c) \setminus \{1\}$ *we have*

$$\mathbf{P}(H_Z(x) \geq \eta - 4\gamma \text{ for all } x \in (0,1) \setminus S_\eta) = 1.$$

Lemma 7.3. *For every* $\eta \in (\eta_c, \overline{\eta}_c) \setminus \{1\}$ *we have*

$$\dim(S_\eta) \leq (1+\beta)(\eta - \eta_c), \quad \mathbf{P} - \text{a.s.}$$

The aim of the Lemma 7.2 is to show that for any *fixed* $\eta \in (\eta_c, \overline{\eta}_c)$ and any $\varepsilon > 0$ sufficiently small, outside the set $S_{\eta+4\gamma-\varepsilon}$ the Hölder exponent is *larger* than $\eta + \varepsilon$, and hence $\tilde{\mathscr{E}}_{Z,\eta} \subset S_{\eta+4\gamma-\varepsilon}$. Therefore Lemma 7.3 gives immediately the upper bound on dimension of $\tilde{\mathscr{E}}_{Z,\eta}$ for any fixed $\eta \in (\eta_c, \overline{\eta}_c)$. Now we will show how these results imply.

Proof of Proposition 7.1. By the definitions of $\mathscr{E}_{Z,\eta}$ and $\tilde{\mathscr{E}}_{Z,\eta}$, we have $\mathscr{E}_{Z,\eta} \subset \tilde{\mathscr{E}}_{Z,\eta}$ for any η and hence the first inequality in (7.1) is obvious. The remaining part of the proof is devoted to deriving the second inequality in (7.1). First we note that this inequality is obvious for $\eta = \overline{\eta}_c$. Now, it follows easily from Lemma 7.2 that $\tilde{\mathscr{E}}_{Z,\eta} \subset S_{\eta+4\gamma+\varepsilon}$ for every $\eta \in (\eta_c, \overline{\eta}_c) \setminus \{1\}$ and every $\varepsilon > 0$ sufficiently small. Therefore,

$$\dim(\tilde{\mathscr{E}}_{Z,\eta}) \leq \lim_{\varepsilon \to 0} \dim(S_{\eta+4\gamma+\varepsilon}).$$

Using Lemma 7.3 we then get

$$\dim(\tilde{\mathscr{E}}_{Z,\eta}) \leq (1+\beta)(\eta + 4\gamma - \eta_c), \quad \mathbf{P} - \text{a.s.}$$

Since γ can be chosen arbitrary small, we have for every $\eta \in (\eta_c, \overline{\eta}_c) \setminus \{1\}$

$$\dim(\tilde{\mathscr{E}}_{Z,\eta}) \leq (1+\beta)(\eta - \eta_c), \quad \mathbf{P} - \text{a.s.}$$

The validity of this inequality for $\eta = 1$ and $\eta = \eta_c$ follows from the monotonicity in η of the sets $\tilde{\mathscr{E}}_{Z,\eta}$. Thus, putting everything together, we have for every $\eta \in [\eta_c, \overline{\eta}_c]$,

$$\dim(\tilde{\mathscr{E}}_{Z,\eta}) \leq (1+\beta)(\eta - \eta_c), \quad \mathbf{P} - \text{a.s.}$$

This bound is not uniform in η, that is, the set of ω, on which this bound can fail, may depend on η. However, this bound is uniform on the set of all rational values of $\eta \in [\eta_c, \overline{\eta}_c]$, that is,

$$\dim(\tilde{\mathscr{E}}_{Z,\eta}) \leq (1+\beta)(\eta - \eta_c), \quad \eta \in [\eta_c, \overline{\eta}_c] \cap \mathbb{Q}, \quad \mathbf{P} - \text{a.s.},$$

where $\mathbb{Q}$ is the set of rational numbers. By monotonicity in η of the sets $\tilde{\mathscr{E}}_{Z,\eta}$, we have (7.1).

$\square$

Let $\varepsilon \in (0, \eta_c/2)$ be arbitrarily small. We introduce a new "good" event A_2^ε which will be frequently used throughout the proofs. This event plays the same role as A_1^ε in the proof of Theorem 1.3 in Chapter 6.

Recall that, by Theorem 1.2 in [18], $x \mapsto X_t(x)$ is **P**-a.s. Hölder continuous with any exponent less than η_c. Hence, for every ε small enough, we can define a constant $C_{(7.2)} = C_{(7.2)}(\varepsilon)$ such that

$$\mathbf{P}\left(\sup_{x_1,x_2 \in (0,1), x_1 \neq x_2} \frac{|X_t(x_1) - X_t(x_2)|}{|x_1 - x_2|^{\eta_c - \varepsilon}} \leq C_{(7.2)} \right) \geq 1 - \varepsilon. \tag{7.2}$$

Now we are ready to define

$$A_2^\varepsilon := \left\{ \sup_{x_1,x_2 \in (0,1), x_1 \neq x_2} \frac{|X_t(x_1) - X_t(x_2)|}{|x_1 - x_2|^{\eta_c - \varepsilon}} \leq C_{(7.2)} \right\} \tag{7.3}$$
$$\cap \{|\Delta X_s| \leq C_{(4.5)}(t-s)^{(1+\beta)^{-1} - \gamma} \text{ for all } s < t\} \cap \{V(B_2(0)) \leq c_\varepsilon\},$$

where $V(\cdot)$ is defined in Lemma 3.5 and c_ε is chosen as in (6.4). Note that A_2^ε differs from A_1^ε only by the first set in their definitions.

Clearly by (4.5), (6.4), and (7.2) we have

$$\mathbf{P}(A_2^\varepsilon) \geq 1 - 3\varepsilon. \tag{7.4}$$

See (3.4) in [18] for the analogous definition.

Now we are ready to give the proof of Lemma 7.3.

Proof of Lemma 7.3. To every jump (s, y, r) of the measure $\mathcal{N}$ (in what follows in the book we will usually call them simply "jumps") with

$$(s, y, r) \in D_{j,n} := [t - 2^{-j}, t - 2^{-j-1}) \times (0,1) \times [2^{-n-1}, 2^{-n})$$

we assign the ball

$$B^{(s,y,r)} := \text{ open ball with center at } y \text{ and radius } \left(\frac{2^{-n}}{(2^{-j-1})^{\frac{1}{1+\beta} - \gamma}} \right)^{1/(\eta - \eta_c)}. \tag{7.5}$$

Define $n_0(j) := j[\frac{1}{1+\beta} - \frac{\gamma}{4}]$. It follows from (4.5) and (7.3) that, on A_2^ε, there are no jumps bigger than $2^{-n_0(j)}$ in the time interval $[t - 2^{-j}, t - 2^{-j-1})$.

It is easy to see that every point from S_η is contained in infinitely many balls $B^{(s,y,r)}$. Therefore, for every $J \geq 1$, the set

$$\bigcup_{j \geq J, n \geq 1} \bigcup_{(s,y,r) \in D_{j,n}} B^{(s,y,r)}$$

covers S_η. From (4.5) and (7.3) we conclude that, on A_2^ε, there are no jumps bigger than $c_{(4.5)}2^{-(j+1)(\frac{1}{1+\beta}-\gamma)}$ in the time interval $s \in [t - 2^{-j}, t - 2^{-j-1})$ for any $j \geq 1$. Define $n_0(j) := j[\frac{1}{1+\beta} - \frac{\gamma}{4}]$. Clearly, there exists J_0 such that for all $j \geq J_0$ there are no jumps bigger than $2^{-n_0(j)}$ in the time interval $[t - 2^{-j}, t - 2^{-j-1})$. Hence, for every $J \geq J_0$, the set

$$S_\eta(J) := \bigcup_{j \geq J, n \geq n_0(j)} \bigcup_{(s,y,r) \in D_{j,n}} B^{(s,y,r)}$$

covers S_η for every $\omega \in A_2^\varepsilon$.

It follows from the formula for the compensator that, on the event $\{\sup_{s \leq t} X_s((0,1)) \leq N\}$, the intensity of jumps with $(s,y,r) \in D_{j,n}$ is bounded by

$$N2^{-j-1}\int_{2^{-n-1}}^{2^{-n}} c_\beta r^{-2-\beta} dr = \frac{Nc_\beta(2^{1+\beta}-1)}{2(1+\beta)}2^{n(1+\beta)-j} =: \lambda_{j,n}.$$

Therefore, the intensity of jumps with $(s,y,r) \in \cup_{n=n_0(j)}^{n_1(j)} D_{j,n} =: \tilde{D}_j$, where $n_1(j) = j[\frac{1}{1+\beta} + \frac{\gamma}{4}]$, is bounded by

$$\sum_{n=n_0(j)}^{n_1(j)} \lambda_{j,n} \leq \frac{Nc_\beta 2^\beta}{(\beta+1)}2^{j(1+\beta)\gamma/4} =: \Lambda_j.$$

The number of such jumps does not exceed $2\Lambda_j$ with the probability $1 - e^{-(1-2\log 2)\Lambda_j}$. This is immediate from the exponential Chebyshev inequality applied to Poisson distributed random variables. Analogously, the number of jumps with $(s,y,r) \in D_{j,n}$ does not exceed $2\lambda_{j,n}$ with the probability at least $1 - e^{-(1-2\log 2)\lambda_{j,n}}$. Since

$$\sum_j \left(e^{-(1-2\log 2)\Lambda_j} + \sum_{n=n_1(j)}^\infty e^{-(1-2\log 2)\lambda_{j,n}}\right) < \infty,$$

we conclude, applying the Borel-Cantelli lemma, that, for almost every ω from the set $A_2^\varepsilon \cap \{\sup_{s \leq t} X_s((0,1)) \leq N\}$, there exists $J(\omega)$ such that for all $j \geq J(\omega)$ and $n \geq n_1(j)$, the numbers of jumps in $\tilde{D}_j$ and in $D_{j,n}$ are bounded by $2\Lambda_j$ and $2\lambda_{j,n}$, respectively.

The radius of every ball corresponding to the jump in $\tilde{D}_j$ is bounded by $r_j := C2^{-\frac{3\gamma}{4(\eta-\eta_c)}j}$. Thus, one can easily see that

$$\sum_{j=1}^\infty \left(2\Lambda_j r_j^\theta + \sum_{n=n_1(j)}^\infty 2\lambda_{j,n}\left(\frac{2^{-n}}{(2^{-j-1})^{\frac{1}{1+\beta}-\gamma}}\right)^{\theta/(\eta-\eta_c)}\right) < \infty$$

for every $\theta > (1+\beta)(\eta - \eta_c)$. This yields

$$\dim(S_\eta) \leq (1+\beta)(\eta - \eta_c) \tag{7.6}$$

for almost every $\omega \in A_2^\varepsilon \cap \{\sup_{s \le t} X_s((0,1)) \le N\}$. Since $\sup_{s \le t} X_s((0,1))$ is finite with probability one, then (7.6) holds for almost every $\omega \in A_2^\varepsilon$. Since we can choose ε arbitrarily small, the result follows by (7.4). $\qquad\qquad\qquad\qquad\qquad\qquad\qquad\square$

The proof of Lemma 7.2 is omitted since it goes similarly to the proof of Theorem 1.3(a) and all the modifications come from the necessity of dealing with numerous random points. For details of the proof of Lemma 7.2, we refer the interested reader to the proof of Lemma 3.2 in [38].

7.2 Lower bound for the Hausdorff dimension

The aim of this section is to describe the main steps in the proof of the following proposition. The full proof of it is given in the Section 4 of [38].

Proposition 7.4. *For every* $\eta \in (\eta_c, \overline{\eta}_c) \setminus \{1\}$,

$$\dim(\mathscr{E}_{Z,\eta}) \ge (1+\beta)(\eta - \eta_c), \quad \mathbf{P} - \text{a.s. on } \{X_t((0,1)) > 0\}.$$

Remark 7.5. Clearly the above proposition together with Proposition 7.1 finishes the proof of Theorem 1.5.

The proof of the lower bound is much more involved than the proof of the upper one. Let us give short description of the strategy. First we state two lemmas that give some uniform estimates on "masses" of X_s of dyadic intervals at times s close to t. These lemmas imply that $X_s(dx)$ for times s close to t is very close to Lebesgue measure with the density bounded from above and away of zero. This is very helpful for constructing a set $J_{\eta,1}$ with $\dim(J_{\eta,1}) \ge (\beta+1)(\eta - \eta_c)$, on which we show existence of "big" jumps of X that occur close to time t. These jumps are "encoded" in the jumps of the auxiliary processes $L_{n,l,r}^+$ and they, in fact, "may" destroy the Hölder continuity of $X_t(\cdot)$ on $J_{\eta,1}$ for any index greater or equal to η (see Proposition 7.10 and the proof of Proposition 7.4).

In the next two lemmas we give some bounds for $X_s(I_k^{(n)})$, where

$$I_k^{(n)} := [k2^{-n}, (k+1)2^{-n}).$$

In what follows, fix some

$$m > 3/2, \qquad\qquad\qquad\qquad\qquad\qquad\qquad (7.7)$$

and let $\theta \in (0,1)$ be arbitrarily small. Define

$$O_n := \{\omega : \text{there exists } k \in [0, 2^n - 1] \text{ such that}$$

$$\sup_{s \in (t-2^{-2n}n^{4m/3}, t)} X_s(I_k^{(n)}) \ge 2^{-n}n^{4m/3}\}$$

and

$$B_n = B_n(\theta) := \{\omega : \text{ there exists } k \in [0, 2^n - 1] \text{ such that}$$

$$I_k^{(n)} \cap \{x : X_t(x) \geq \theta\} \neq \emptyset$$

$$\text{and} \quad \inf_{s \in (t - 2^{-2n} n^{-2m}, t)} X_s(I_k^{(n)}) \leq 2^{-n} n^{-2m} \}.$$

Lemma 7.6. *There exists a constant C such that*

$$\mathbf{P}(O_n) \leq C n^{-2m/3}, \ n \geq 1.$$

Recall the definition of A_2^ε given in (7.3).

Lemma 7.7. *There exists a constant $C = C(m)$ such that, for every $\theta \in (0, 1)$,*

$$\mathbf{P}(B_n(\theta) \cap A_2^\varepsilon) \leq C\theta^{-1} n^{-\alpha m/3}, \ n \geq \tilde{n}(\theta),$$

for some $\tilde{n}(\theta)$ sufficiently large.

The proofs of the above lemmas are technical and hence we omit them. Let us just mention that the proof of Lemma 7.6 is an almost word-by-word repetition of the proof of Lemma 5.5 in [18], and for the proof of Lemma 7.7 we refer the reader to the proof of Lemma 6.7 in [38].

7.2.1 Analysis of the set of jumps which destroy the Hölder continuity

In this subsection we construct a set $J_{\eta,1}$ such that its Hausdorff dimension is bounded from below by $(\beta + 1)(\eta - \eta_c)$ and in the vicinity of each $x \in J_{\eta,1}$ there are jumps of X which destroy the Hölder continuity at x for any index greater than η.

We first introduce $J_{\eta,1}$ and prove the lower bound for its dimension. Set

$$q := \frac{5m}{(\beta + 1)(\eta - \eta_c)}$$

and define

$$A_k^{(n)} := \left\{ \Delta X_s(I_{k-2n^q-2}^{(n)}) \geq 2^{-(\eta+1)n} \right.$$

$$\left. \text{for some } s \in [t - 2^{-2n} n^{-2m}, t - 2^{-2(n+1)}(n+1)^{-2m}) \right\},$$

$$J_{k,r}^{(n)} := \left[\frac{k}{2^n} - (n^q 2^{-n})^r, \frac{k+1}{2^n} + (n^q 2^{-n})^r \right].$$

Let us introduce the following notation. For a Borel set B and an event E define a random set

$$B1_E(\omega) := \begin{cases} B, & \omega \in E, \\ \emptyset, & \omega \notin E. \end{cases}$$

Now we are ready to define random sets

$$J_{\eta,r} := \limsup_{n \to \infty} \bigcup_{k=2n^q+2}^{2^n-1} J_{k,r}^{(n)} 1_{A_k^{(n)}}, \quad r > 0.$$

As we have mentioned already we are interested in getting the lower bound on Hausdorff dimension of $J_{\eta,1}$. The standard procedure for this is as follows. First shows that a bit "inflated" set $J_{\eta,r}$, for certain $r \in (0,1)$, contains open intervals. This would imply a lower bound r on the Hausdorff dimension of $J_{\eta,1}$ (see Lemma 7.8 and Theorem 2 from [28] where a similar strategy was implemented). Thus to get a sharper bound on Hausdorff dimension of $J_{\eta,1}$ one should try to take r as large as possible. In the next lemma we show that, in fact, one can choose $r = (\beta+1)(\eta - \eta_c)$.

Lemma 7.8. *On the event A_2^ε,*

$$\{x \in (0,1) : X_t(x) \geq \theta\} \subseteq J_{\eta,(\beta+1)(\eta-\eta_c)}, \quad \mathbf{P} - \text{a.s.}$$

for every $\theta \in (0,1)$.

Proof. Fix an arbitrary $\theta \in (0,1)$. We estimate the probability of the event $E_n \cap A_2^\varepsilon$, where

$$E_n := \left\{ \omega : \{x \in (0,1) : X_t(x) \geq \theta\} \subseteq \bigcup_{k=2n^q+2}^{2^n-1} J_{k,(\beta+1)(\eta-\eta_c)}^{(n)} 1_{A_k^{(n)}} \right\}.$$

To prove the lemma it is enough to show that the sequence $\mathbf{P}(E_n^c \cap A_2^\varepsilon)$ is summable. It follows from Lemma 7.7 that, for all $n \geq \tilde{n}(\theta)$,

$$\mathbf{P}(E_n^c \cap A_2^\varepsilon) \leq \mathbf{P}(E_n^c \cap B_n \cap A_2^\varepsilon) + \mathbf{P}(E_n^c \cap B_n^c \cap A_2^\varepsilon)$$
$$\leq C\theta^{-1} n^{-2m/3} + \mathbf{P}(E_n^c \cap B_n^c \cap A_2^\varepsilon). \tag{7.8}$$

For any $k = 0, \ldots, 2^n - 1$, the compensator measure $\widehat{N}(dr, dy, ds)$ of the random measure $\mathcal{N}(dr, dy, ds)$ (the jump measure for X — see Lemma 2.1), on

$$\mathcal{I}_1^{(n)} \times I_k^{(n)} \times \mathcal{I}_2^{(n)}$$
$$:= [2^{-(\eta+1)n}, \infty) \times I_k^{(n)} \times [t - 2^{-2n} n^{-2m}, t - 2^{-2(n+1)}(n+1)^{-2m}),$$

is given by the formula

$$1\{(r,y,s) \in \mathcal{I}_1^{(n)} \times I_k^{(n)} \times \mathcal{I}_2^{(n)}\} n(dr) X_s(dy) ds. \tag{7.9}$$

If
$$k \in K_\theta := \{l : I_l^{(n)} \cap \{x \in (0,1) : X_t(x) \geq \theta\} \neq \emptyset\},$$

then, by the definition of B_n, we have

$$X_s(I_k^{(n)}) \geq 2^{-n} n^{-2m}, \quad \text{for } s \in \mathscr{I}_2^{(n)}, \text{ on the event } A_2^\varepsilon \cap B_n^c. \tag{7.10}$$

Define the measure $\widehat{\Gamma}(dr, dy, ds)$ on $\mathbb{R}_+ \times (0,1) \times \mathbb{R}_+$, as follows:

$$\widehat{\Gamma}(dr, dy, ds) := n(dr) n^{-2m} dy ds. \tag{7.11}$$

Then, by (7.9) and (7.10), on $A_2^\varepsilon \cap B_n^c$, and on the set

$$\mathscr{I}_1^{(n)} \times \{y \in (0,1) : X_t(y) \geq \theta\} \times \mathscr{I}_2^{(n)}$$

we have the following bound:

$$\widehat{\Gamma}(dr, I_k^{(n)}, \mathscr{I}_2^{(n)}) \leq \widehat{\mathscr{N}}(dr, I_k^{(n)}, \mathscr{I}_2^{(n)}), k \in K_\theta.$$

By standard arguments it is easy to construct the Poisson point process $\Gamma(dr, dx, ds)$ on $\mathbb{R}_+ \times (0,1) \times \mathbb{R}_+$ with intensity measure $\widehat{\Gamma}$ given by (7.11) on the whole space $\mathbb{R}_+ \times (0,1) \times \mathbb{R}_+$ such that on $A_2^\varepsilon \cap B_n^c$,

$$\Gamma(dr, I_k^{(n)}, \mathscr{I}_2^{(n)}) \leq \mathscr{N}(dr, I_k^{(n)}, \mathscr{I}_2^{(n)})$$

for $r \in \mathscr{I}_1^{(n)}$ and $k \in K_\theta$.

Now, define

$$\xi_k^{(n)} = 1_{\left\{\Gamma\left(\mathscr{I}_1^{(n)} \times I_{k-2n^q-2}^{(n)} \times \mathscr{I}_2^{(n)}\right) \geq 1\right\}}, \quad k \geq 2n^q + 2.$$

Clearly, on $A_2^\varepsilon \cap B_n^c$ and for k such that $k - 2n^q - 2 \in K_\theta$,

$$\xi_k^{(n)} \leq 1_{A_k^{(n)}}.$$

Moreover, by construction $\{\xi_k^{(n)}\}_{k=2n^q+2}^{2^n+2n^q+1}$ is a collection of independent identically distributed Bernoulli random variables with success probabilities

$$p^{(n)} := \widehat{\Gamma}\left(\mathscr{I}_1^{(n)} \times I_{k-2n^q-2}^{(n)} \times \mathscr{I}_2^{(n)}\right)$$
$$= C 2^{(\eta - \eta_c)(1+\beta)n - n} n^{-4m}.$$

From the above coupling with the Poisson point process Γ, it is easy to see that

$$\mathbf{P}(E_n^c \cap B_n^c \cap A_2^\varepsilon) \leq \mathbf{P}(\tilde{E}_n^c), \tag{7.12}$$

where

$$\tilde{E}_n := \left\{ (0,1) \subseteq \bigcup_{k=2n^q+2}^{2^n+2n^q+1} J_k^{(n)} 1_{\{\xi_k^{(n)}=1\}} \right\}.$$

Let $L(n)$ denote the length of the longest run of zeros in the sequence $\{\xi_k^{(n)}\}_{k=2n^q+2}^{2^n+2n^q+1}$. Clearly,

$$\mathbf{P}(\tilde{E}_n^c) \le \mathbf{P}(L^{(n)} \ge 2^{n-(\beta+1)(\eta-\eta_c)n} n^{5m})$$

and it is also obvious that

$$\mathbf{P}(L^{(n)} \ge j) \le 2^n p^{(n)} (1-p^{(n)})^j, \forall j \ge 1.$$

Use this with the fact that, by (7.7), $m > 1$, to get that

$$\mathbf{P}(\tilde{E}_n^c) \le \exp\left\{ -\frac{1}{2} n^m \right\} \tag{7.13}$$

for all n sufficiently large. Combining (7.8), (7.12), and (7.13), we conclude that the sequence $\mathbf{P}(E_n^c \cap A_2^\varepsilon)$ is summable. Applying Borel-Cantelli, we complete the proof of the lemma. $\qquad\square$

Define

$$h_\eta(x) := x^{(\beta+1)(\eta-\eta_c)} \log^2 \frac{1}{x}$$

and

$$\mathscr{H}_\eta(A) := \liminf_{\varepsilon \to 0} \left\{ \sum_{j=1}^{\infty} h_\eta(|I_j|), A \in \bigcup_{j=1}^{\infty} I_j \text{ and } |I_j| \le \varepsilon \right\}.$$

Combining Lemma 7.8 and Theorem 2 from [28], one can easily get

Corollary 7.9. *On the event* $A_2^\varepsilon \cap \{X_t((0,1)) > 0\}$,

$$\mathscr{H}_\eta(J_{\eta,1}) > 0, \quad \mathbf{P}-a.s.$$

and, consequently, on $A_2^\varepsilon \cap \{X_t((0,1)) > 0\}$,

$$\dim(J_{\eta,1}) \ge (\beta+1)(\eta-\eta_c), \quad \mathbf{P}-a.s.$$

Proof. Fix any $\theta \in (0,1)$. If $\omega \in A_2^\varepsilon$ is such that $B_\theta := \{x \in (0,1) : X_t(x) \ge \theta\}$ is not empty, then by the local Hölder continuity of $X_t(\cdot)$ there exists an open interval $(x_1(\omega), x_2(\omega)) \subset B_{\theta/2}$. Moreover, in view of Lemma 7.8,

$$(x_1(\omega), x_2(\omega)) \subset J_{\eta,(\beta+1)(\eta-\eta_c)}(\omega), \quad \mathbf{P}-a.s.$$

on the event $A_2^\varepsilon \cap \{B_\theta \text{ is not empty}\}$. Thus, we may apply Theorem 2 from [28] to the set $(x_1(\omega), x_2(\omega))$, which gives

$$\mathscr{H}_\eta((x_1(\omega), x_2(\omega)) \cap J_{\eta,1}) > 0, \quad \mathbf{P}-a.s.$$

on the event $A_2^\varepsilon \cap \{B_\theta$ is not empty$\}$. Thus,

$$\dim((x_1(\omega), x_2(\omega)) \cap J_{\eta,1}) \geq (\beta + 1)(\eta - \eta_c), \quad \mathbf{P}-\text{a.s.}$$

on the event $A_2^\varepsilon \cap \{B_\theta$ is not empty$\}$. Due to the monotonicity of $\mathscr{H}_\eta(\cdot)$ and $\dim(\cdot)$, we conclude that $\mathscr{H}_\eta(J_{\eta,1}) > 0$ and $\dim(J_{\eta,1}) \geq (\beta + 1)(\eta - \eta_c)$, $\mathbf{P}$-a.s. on the event $A_2^\varepsilon \cap \{B_\theta$ is not empty$\}$. Noting that $1_{\{B_\theta \text{ is not empty}\}} \uparrow 1_{\{X_t(0,1) > 0\}}$ as $\theta \downarrow 0$, $\mathbf{P}$-a.s., we complete the proof. □

Now we turn to the second part of the present subsection. By construction of $J_{\eta,1}$ we know that to the left of every point $x \in J_{\eta,1}$ there exist big jumps of X at time s "close" to t: such jumps are defined by the events $A_k^{(n)}$. We would like to show that these jumps will result in destroying the Hölder continuity of any index greater than η at the point x. To this end, we will introduce auxiliary processes $L_{n,y,x}^{\pm}$ that are indexed by points (y,x) on a grid *finer* than $\{k2^{-n}, k = 0, 1, \ldots\}$. That is, take some integer $Q > 1$ (note that eventually Q will be chosen large enough, depending on η). Define

$$Z_s^\eta(x_1, x_2) := \int_{0^-}^s \int_{\mathbb{R}} M(d(u,y)) \, p_{t-u}^\eta(x_1 - y, x_2 - y), \ s \in [0,t],$$

where

$$p_s^\eta(x,y) := \begin{cases} p_s(x) - p_s(y), & \text{if } \eta \leq 1, \\ p_s(x) - p_s(y) - (x-y)\frac{\partial p_s(y)}{\partial y}, & \text{if } \eta \in (1, \bar{\eta}_c). \end{cases}$$

According to (2.19) and (2.21), for every $x, y \in 2^{-Qn}\mathbb{Z}$, there exist spectrally positive $(1+\beta)$-stable processes $L_{n,y,x}^{\pm}$ such that

$$Z_s^\eta(y,x) = L_{n,y,x}^+(T_+^{n,y,x}(s)) - L_{n,y,x}^-(T_-^{n,y,x}(s)) \tag{7.14}$$
$$=: \mathbf{L}_{n,y,x}^+ - \mathbf{L}_{n,y,x}^-,$$

where

$$T_\pm^{n,y,x}(s) = \int_0^s du \int_{\mathbb{R}} X_u(dz) \left(\left(p_{t-u}^\eta(y - z, x - z) \right)^{\pm} \right)^{1+\beta}, \ s \leq t.$$

In what follows let $[z]$ denote the integer part of z for $z \in \mathbb{R}$. The crucial ingredient for the proof of the lower bound is the following proposition:

Proposition 7.10. *Fix arbitrary $\eta \in (\eta_c, \bar{\eta}_c) \setminus \{1\}$ and $Q > 1$. For $\mathbf{P}$-a.s. ω on A_2^ε, there exists a set $\mathbf{G}_\eta \in [0,1]$ with*

$$\dim(\mathbf{G}_\eta) < (\beta + 1)(\eta - \eta_c) \tag{7.15}$$

such that the following holds. For $\mathbf{P}$-a.s. ω on A_2^ε, for every $x \in J_{\eta,1} \setminus \mathbf{G}_\eta$, there exist a (random) sequences

$$n_j = n_j(x), \ j \geq 1,$$

and

$$(x_{n_j}, y_{n_j}) = (x_{n_j}(x), y_{n_j}(x)), \quad j \geq 1$$

with

$$x_{n_j} = 2^{-Qn_j}[2^{Qn_j}x], \quad j \geq 1,$$

$$|y_{n_j} - x_{n_j}| \leq Cn_j^q 2^{-n_j}, \quad j \geq 1,$$

such that

$$\mathbf{L}^+_{n_j, y_{n_j}, x_{n_j}} \geq n_j^m 2^{-\eta n_j}, \quad \mathbf{L}^-_{n_j, y_{n_j}, x_{n_j}} \leq 2^{-(\eta n_j - 1)},$$

for all n_j sufficiently large.

Note that in the above proposition we do not give precise definition of y_{n_j}, but it is chosen in a way that it is "close" to the spatial position of a "big" jump that is supposed to destroy the Hölder continuity at x. As for the point x_{n_j}, it is chosen to be "close" to x itself.

Similarly to Proposition 6.3, Proposition 7.10 deals with possible compensation effects. In contrast to the case of fixed points considered in Proposition 6.3, we cannot show that the compensation does not happen. But we derive an upper bound for the Hausdorff dimension of the set $\mathbf{G}_\eta$, on which such a compensation may occur. The dimension of $\mathbf{G}_\eta$ turns to be strictly smaller than $(\beta + 1)(\eta - \eta_c)$, see (7.15).

The proof of Proposition 7.10 is rather technical and uses the same ideas as the proof of Proposition 6.3. The major additional difficulty comes from the need to consider random sets. To overcome it we use Borel-Cantelli arguments. We refer the reader to Section 4.3 in [38] where the proofs of results leading to Proposition 7.10 are given. Now we will explain how Proposition 7.10 implies the proof of Proposition 7.4 for the case of $\eta < 1$ (the proof for the case of $\eta > 1$ goes along the similar lines, see Section 4.4 in [38]).

Proof of Proposition 7.4 for $\eta < 1$. Fix arbitrary $\eta \in (\eta_c, \min(\bar{\eta}_c, 1))$. Also fix

$$Q = \left[4\frac{\eta}{\eta_c} + 2\right],$$

where as usual $[z]$ denotes the integer part of z. Let $\mathbf{G}_\eta$ be as in Proposition 7.10. If x is an arbitrary point in $J_{\eta,1} \setminus \mathbf{G}_\eta$, then let $\{n_j(x)\}_{j \geq 1}$ and $\{(x_{n_j}(x), y_{n_j}(x))\}_{j \geq 1}$ be the sequences constructed in Proposition 7.10. Then Proposition 7.10 implies that,

$$\liminf_{j \to \infty} 2^{(\eta + \delta)n_j} \left|Z_t^\eta \left(y_{n_j}(x), x_{n_j}(x)\right)\right| = \infty, \quad \forall x \in J_{\eta,1} \setminus \mathbf{G}_\eta, \ \mathbf{P} - \text{a.s. on } A_2^\varepsilon. \quad (7.16)$$

for any $\delta > 0$. Recall that, $X_t(\cdot)$ and $Z_t(\cdot)$ are Hölder continuous with any exponent less than η_c at every point of $(0,1)$. Therefore, recalling that $Q > 4\frac{\eta}{\eta_c}$, we have

$$\lim_{j \to \infty} \sup_{x \in (0,1)} 2^{(\eta+\delta)n_j} |Z_t^\eta(x, x_{n_j}(x))|$$

$$= \lim_{j \to \infty} C(\omega) 2^{-\frac{1}{2}\varrho \eta_c n_j} 2^{(\eta+\delta)n_j} = 0, \quad \mathbf{P} - \text{a.s. on } A_2^\varepsilon. \tag{7.17}$$

Therefore, for any x in $J_{\eta,1} \setminus \mathbf{G}_\eta$, we have

$$|Z_t^\eta(y_{n_j}(x), x)| \geq |Z_t^\eta(y_{n_j}(x), x_{n_j}(x))| - |Z_t^\eta(x_{n_j}(x), x)|, \quad j \geq 1. \tag{7.18}$$

Therefore, combining (7.16), (7.17), and (7.18) we conclude that

$$H_Z(x) \leq \eta, \quad \text{for all } x \in J_{\eta,1} \setminus \mathbf{G}_\eta, \quad \mathbf{P} - \text{a.s. on } A_2^\varepsilon. \tag{7.19}$$

We know, by Lemma 7.2, that

$$H_Z(x) \geq \eta - 2\gamma - 2\rho \quad \text{for all } x \in (0,1) \setminus S_{\eta-2\rho}, \quad \mathbf{P} - \text{a.s.},$$

This and (7.19) imply that on $A_2^\varepsilon, \mathbf{P} - \text{a.s.}$,

$$\eta - 2\gamma - 2\rho \leq H_Z(x) \leq \eta \quad \text{for all } x \in (J_{\eta,1} \setminus S_{\eta-2\rho}) \setminus \mathbf{G}_\eta. \tag{7.20}$$

It follows easily from Lemma 7.3, Corollary 7.9 and Lemma 2.9 that on A_2^ε

$$\dim\left((J_{\eta,1} \setminus S_{\eta-2\rho}) \setminus \mathbf{G}_\eta\right) \geq (\beta+1)(\eta - \eta_c), \quad \mathbf{P} - \text{a.s.}$$

Thus, by (7.20),

$$\dim\{x : H_Z(x) \leq \eta\} \geq (\beta+1)(\eta - \eta_c), \quad \text{on } A_2^\varepsilon, \mathbf{P} - \text{a.s..}$$

It is clear that

$$\{x : H_Z(x) = \eta\} \cup \bigcup_{n=n_0}^{\infty} \{x : H_Z(x) \in (\eta - n^{-1}, \eta - (n+1)^{-1}]\}$$

$$= \{x : \eta - n_0^{-1} \leq H_Z(x) \leq \eta\}.$$

Consequently,

$$\mathscr{H}_\eta(\{x : \eta - n_0^{-1} \leq H_Z(x) \leq \eta\})$$
$$= \mathscr{H}_\eta(\{x : H_Z(x) = \eta\})$$
$$+ \sum_{n=n_0}^{\infty} \mathscr{H}_\eta(\{x : H_Z(x) \in (\eta - n^{-1}, \eta - (n+1)^{-1}]\}).$$

Since the dimensions of $S_{\eta-2\rho}$ and $\mathbf{G}_\eta$ are smaller than η, the $\mathscr{H}_\eta$-measure of these sets equals zero. Applying Corollary 7.9, we then conclude that on A_2^ε

$$\mathscr{H}_\eta((J_{\eta,1} \setminus S_{\eta-2\rho}) \setminus \mathbf{G}_\eta) > 0, \mathbf{P} - \text{a.s.}$$

And in view of (7.20), $\mathscr{H}_\eta(\{x : \eta - n_0^{-1} \le H_Z(x) \le \eta\}) > 0$. Furthermore, it follows from Proposition 7.1, that dimension of the set $\{x : H_Z(x) \in (\eta - n^{-1}, \eta - (n+1)^{-1}]\}$ is bounded from above by $(\beta + 1)(\eta - (n+1)^{-1} - \eta_c)$. Hence, the definition of $\mathscr{H}_\eta$ immediately yields

$$\mathscr{H}_\eta(\{x : H_Z(x) \in (\eta - n^{-1}, \eta - (n+1)^{-1}]\}) = 0, \quad \text{on } A_2^\varepsilon, \ \mathbf{P} - \text{a.s.},$$

for all $n \ge n_0$. As a result we have

$$\mathscr{H}_\eta(\{x : H_Z(x) = \eta\}) > 0 \ \mathbf{P} - \text{a.s. on } A_2^\varepsilon. \tag{7.21}$$

Since $\varepsilon > 0$ was arbitrary, this implies that (7.21) is satisfied on the whole probability space $\mathbf{P}$-a.s. From this we get that

$$\dim\{x : H_Z(x) = \eta\} \ge (\beta + 1)(\eta - \eta_c), \ \mathbf{P} - \text{a.s.}$$

$\square$

Appendix A
Estimates for the transition kernel of the one-dimensional Brownian motion

We start with the following estimates for p_t which are taken from Rosen [43].

Lemma A.1. *Let $d = 1$. For each $\delta \in (0,1]$ there exists a constant C such that*

$$\left| p_t(x) - p_t(y) \right| \leq C \frac{|x-y|^\delta}{t^{\delta/2}} \left(p_t(x/2) + p_t(y/2) \right), \tag{A.1}$$

$$\left| \frac{\partial p_t(x)}{\partial x} \right| \leq C t^{-1/2} p_t(x/2), \tag{A.2}$$

$$\left| \frac{\partial p_t(x)}{\partial x} - \frac{\partial p_t(y)}{\partial y} \right| \leq C \frac{|x-y|^\delta}{t^{(1+\delta)/2}} \left(p_t(x/2) + p_t(y/2) \right), \tag{A.3}$$

$$\left| p_t(x) - p_t(y) - (x-y) \frac{\partial p_t(y)}{\partial y} \right| \leq C \frac{|x-y|^{1+\delta}}{t^{(1+\delta)/2}} \left(p_t(x/2) + p_t(y/2) \right) \tag{A.4}$$

for all $t > 0$ and $x, y \in \mathbb{R}$.

The next lemma is a simple corollary of the previous one.

Lemma A.2. *Let $d = 1$. If $\theta \in (1,3)$ and $\delta \in (0,1]$ satisfy $\delta < (3 - \theta)/\theta$, then*

$$\int_0^t ds \int_\mathbb{R} dy \, p_s(y) \left| p_{t-s}(x_1 - y) - p_{t-s}(x_2 - y) \right|^\theta$$
$$\leq C(1+t)|x_1 - x_2|^{\delta\theta} \left(p_t(x_1/2) + p_t(x_2/2) \right), \quad t > 0, \; x_1, x_2 \in \mathbb{R}.$$

Proof. By Lemma A.1, for every $\delta \in [0,1]$,

$$\left| p_{t-s}(x_1 - y) - p_{t-s}(x_2 - y) \right|^\theta$$
$$\leq C \frac{|x_1 - x_2|^{\delta\theta}}{(t-s)^{\delta\theta/2}} \left(p_{t-s}\big((x_1 - y)/2\big) + p_{t-s}\big((x_2 - y)/2\big) \right)^\theta,$$

© The Author(s) 2016
L. Mytnik, V. Wachtel, *Regularity and Irregularity of Superprocesses with (1 + β)-stable Branching Mechanism*, SpringerBriefs in Probability and Mathematical Statistics, DOI 10.1007/978-3-319-50085-0

$t > s \geq 0$, $x_1, x_2, y \in \mathbb{R}$. Noting that $p_{t-s}(\cdot) \leq C(t-s)^{-1/2}$, we obtain

$$\left| p_{t-s}(x_1 - y) - p_{t-s}(x_2 - y) \right|^\theta \tag{A.5}$$

$$\leq C \frac{|x_1 - x_2|^{\delta\theta}}{(t-s)^{(\delta\theta+\theta-1)/2}} \left(p_{t-s}((x_1 - y)/2) + p_{t-s}((x_2 - y)/2) \right),$$

$t > s \geq 0$, $x_1, x_2, y \in \mathbb{R}$. Therefore,

$$\int_0^t ds \int_{\mathbb{R}} dy\, p_s(y) \left| p_{t-s}(x_1 - y) - p_{t-s}(x_2 - y) \right|^\theta \leq C|x_1 - x_2|^{\delta\theta}$$

$$\times \int_0^t ds\, (t-s)^{-(\delta\theta+\theta-1)/2} \int_{\mathbb{R}} dy\, p_s(y) \left(p_{t-s}((x_1 - y)/2) + p_{t-s}((x_2 - y)/2) \right).$$

By scaling of the kernel p,

$$\int_{\mathbb{R}} dy\, p_s(y) p_{t-s}((x - y)/2) = \frac{1}{2} \int_{\mathbb{R}} dy\, p_{s/4}(y/2) p_{t-s}((x_2 - y)/2)$$

$$= \frac{1}{2} p_{s/4+t-s}(x/2) = \frac{1}{2}(s/4+t-s)^{-1/2} p_1\left((s/4+t-s)^{-1/2}x/2\right)$$

$$\leq t^{-1/2} p_1(t^{-1/2}x/2) = p_t(x/2),$$

since $t/4 \leq s + t/4 - s \leq t$.

As a result we have the inequality

$$\int_0^t ds \int_{\mathbb{R}} dy\, p_s(y) \left| p_{t-s}(x_1 - y) - p_{t-s}(x_2 - y) \right|^\theta$$

$$\leq C|x_1 - x_2|^{\delta\theta} \left(p_t(x_1/2) + p_t(x_2/2) \right) \int_0^t ds\, s^{-(\delta\theta+\theta-1)/2}.$$

Noting that the latter integral is bounded by $C(1+t)$, since $(\delta\theta + \theta - 1) < 2$, we get the desired inequality. □

Appendix B
Probability inequalities for a spectrally positive stable process

Let L be a spectrally positive stable process of index κ with Laplace transform given by

$$\mathbf{E}e^{-\lambda L_t} = e^{t\lambda^\kappa}, \quad \lambda, t \geq 0. \tag{B.1}$$

Let $\Delta L_s := L_s - L_{s-} > 0$ denote the jumps of L.

Lemma B.1. *We have*

$$\mathbf{P}\left(\sup_{0 \leq u \leq t} L_u 1\left\{ \sup_{0 \leq v \leq u} \Delta L_v \leq y \right\} \geq x \right) \leq \left(\frac{Ct}{xy^{\kappa-1}} \right)^{x/y}, \quad t > 0, \ x, y > 0.$$

Proof. Since for $r > 0$ fixed, $\{L_{rt} : t \geq 0\}$ is equal to $r^{1/\kappa}L$ in law, for the proof we may assume that $t = 1$. Let $\{\xi_i : i \geq 1\}$ denote a family of independent copies of L_1. Set

$$W_{ns} := \sum_{1 \leq k \leq ns} \xi_k, \quad L_s^{(n)} := n^{-1/\kappa}W_{ns}, \quad 0 \leq s \leq 1, \ n \geq 1.$$

Denote by $D_{[0,1]}$ the Skorokhod space of càdlàg functions $f : [0,1] \to \mathbb{R}$. For fixed $y > 0$, let $H : D_{[0,1]} \mapsto \mathbb{R}$ be defined by

$$H(f) = \sup_{0 \leq u \leq 1} f(u) 1\left\{ \sup_{0 \leq v \leq u} \Delta f(v) \leq y \right\}, \quad f \in D_{[0,1]}.$$

It is easy to verify that H is continuous on the set $D_{[0,1]} \setminus J_y$, where $J_y := \{f \in D_{[0,1]} : \Delta f(v) = y \text{ for some } v \in [0,1]\}$. Since $\mathbf{P}(L \in J_y) = 0$, from the invariance principle (see, e.g., Gikhman and Skorokhod [22], Theorem 9.6.2) for $L^{(n)}$ we conclude that

$$\mathbf{P}\big(H(L) \geq x\big) = \lim_{n \uparrow \infty} \mathbf{P}\big(H(L^{(n)}) \geq x\big), \quad x > 0.$$

© The Author(s) 2016

L. Mytnik, V. Wachtel, *Regularity and Irregularity of Superprocesses with (1 + β)-stable Branching Mechanism*, SpringerBriefs in Probability and Mathematical Statistics, DOI 10.1007/978-3-319-50085-0

Consequently, the lemma will be proved if we show that

$$\mathbf{P}\Big(\sup_{0 \le u \le 1} W_{nu} 1\{ \max_{1 \le k \le nu} \xi_k \le y n^{1/\kappa} \} \ge x n^{1/\kappa} \Big)$$

$$\le \Big(\frac{C}{xy^{\kappa-1}} \Big)^{x/y}, \quad x, y > 0, \ n \ge 1. \tag{B.2}$$

To this end, for fixed $y', h \ge 0$, we consider the sequence

$$\Lambda_0 := 1, \quad \Lambda_n := e^{h W_n} 1\{ \max_{1 \le k \le n} \xi_k \le y' \}, \quad n \ge 1.$$

It is easy to see that

$$\mathbf{E}\{\Lambda_{n+1} \,|\, \Lambda_n = e^{hu}\} = e^{hu} \mathbf{E}\{e^{h L_1}; L_1 \le y'\} \quad \text{for all } u \in \mathbb{R}$$

and that

$$\mathbf{E}\{\Lambda_{n+1} \,|\, \Lambda_n = 0\} = 0.$$

In other words,

$$\mathbf{E}\{\Lambda_{n+1} \,|\, \Lambda_n\} = \Lambda_n \mathbf{E}\{e^{h L_1}; L_1 \le y'\}. \tag{B.3}$$

This means that $\{\Lambda_n : n \ge 1\}$ is a supermartingale (submartingale) if h satisfies $\mathbf{E}\{e^{h L_1}; L_1 \le y'\} \le 1$ (respectively, $\mathbf{E}\{e^{h L_1}; L_1 \le y'\} \ge 1$). If Λ_n is a submartingale, then by Doob's inequality,

$$\mathbf{P}\Big(\max_{1 \le k \le n} \Lambda_k \ge e^{h x'} \Big) \le e^{-h x'} \mathbf{E}\Lambda_n, \quad x' > 0.$$

But if Λ_n is a supermartingale, then

$$\mathbf{P}\Big(\max_{1 \le k \le n} \Lambda_k \ge e^{h x'} \Big) \le e^{-h x'} \mathbf{E}\Lambda_0 = e^{-h x'}, \quad x' > 0.$$

From these inequalities and (B.3) we get

$$\mathbf{P}\Big(\max_{1 \le k \le n} \Lambda_k \ge e^{h x'} \Big) \le e^{-h x'} \max\Big\{ 1, \big(\mathbf{E}\{e^{h L_1}; L_1 \le y'\} \big)^n \Big\}. \tag{B.4}$$

It was proved by Fuk and Nagaev [21] (see the first formula in the proof of Theorem 4 there) that

$$\mathbf{E}\{e^{h L_1}; L_1 \le y'\} \le 1 + h \mathbf{E}\{L_1; L_1 \le y'\} + \frac{e^{h y'} - 1 - h y'}{(y')^2} V(y'), \ h, y' > 0,$$

where $V(y') := \int_{-\infty}^{y'} \mathbf{P}(L_1 \in du) u^2 > 0$. Noting that the assumption $\mathbf{E}L_1 = 0$ yields that $\mathbf{E}\{L_1; L_1 \le y'\} \le 0$, we obtain

$$\mathbf{E}\{e^{h L_1}; L_1 \le y'\} \le 1 + \frac{e^{h y'} - 1 - h y'}{(y')^2} V(y'), \quad h, y' > 0. \tag{B.5}$$

Now note that

$$\Big\{ \max_{1\le k\le n} W_k 1\{ \max_{1\le i\le k} \xi_i \le y' \} \ge x' \Big\} = \Big\{ \max_{1\le k\le n} e^{hW_k} 1\{ \max_{1\le i\le k} \xi_i \le y' \} \ge e^{hx'} \Big\}$$

$$= \Big\{ \max_{1\le k\le n} \Lambda_k \ge e^{hx'} \Big\}. \tag{B.6}$$

Thus, combining (B.6), (B.5), and (B.4), we get

$$\mathbf{P}\Big(\max_{1\le k\le n} W_k 1\{ \max_{1\le i\le k} \xi_i \le y' \} \ge x' \Big) \le \mathbf{P}\Big(\max_{1\le k\le n} \Lambda_k \ge e^{hx'} \Big)$$

$$\le \exp\Big\{ -hx' + \frac{e^{hy'} - 1 - hy'}{(y')^2} nV(y') \Big\}.$$

Choosing $h := (y')^{-1} \log(1 + x'y'/nV(y'))$, we arrive, after some elementary calculations, at the bound

$$\mathbf{P}\Big(\max_{1\le k\le n} W_k 1\{ \max_{1\le i\le k} \xi_i \le y' \} \ge x' \Big) \le \Big(\frac{enV(y')}{x'y'} \Big)^{x'/y'}, \quad x',y' > 0.$$

Since $\mathbf{P}(L_1 > u) \sim Cu^{-\kappa}$ as $u \uparrow \infty$, we have $V(y') \le C(y')^{2-\kappa}$ for all $y' > 0$. Therefore,

$$\mathbf{P}\Big(\max_{1\le k\le n} W_k 1\{ \max_{1\le i\le k} \xi_i \le y' \} \ge x' \Big) \le \Big(\frac{Cn}{x'(y')^{\kappa-1}} \Big)^{x'/y'}, \quad x',y' > 0. \tag{B.7}$$

Choosing finally $x' = xn^{1/\kappa}$, $y' = yn^{1/\kappa}$, we get (B.2) from (B.7). Thus, the proof of the lemma is complete. $\qquad\square$

Acknowledgements Support by Israel Science Foundation grant 1325/14 is gratefully acknowledged. The authors also thank the referees for their useful comments and suggestions which improved the exposition.

References

1. P. Balança, Fine regularity of Lévy processes and linear (multi)fractional stable motion. Electron. J. Probab. **19**(101), 37 (2014)
2. P. Balança, Some sample path properties of multifractional Brownian motion. Stoch. Process. Appl. **125**(10), 3823–3850 (2015)
3. P. Balança, Uniform multifractal structure of stable trees (2015). ArXiv:1508.00229v1
4. P. Balança, L. Mytnik, Singularities of stable super-Brownian motion (2016). ArXiv:1608.00792v1
5. J. Barral, S. Seuret, Inside singularity sets of random Gibbs measures. J. Stat. Phys. **120**(5-6), 1101–1124 (2005)
6. J. Barral, S. Seuret, Renewal of singularity sets of random self-similar measures. Adv. Appl. Probab. **39**(1), 162–188 (2007)
7. J. Barral, S. Seuret, The singularity spectrum of Lévy processes in multifractal time. Adv. Math. **214**(1), 437–468 (2007)
8. J. Barral, N. Fournier, S. Jaffard, S. Seuret, A pure jump Markov process with a random singularity spectrum. Ann. Probab. **38**(5), 1924–1946 (2010)
9. J. Barral, A. Kupiainen, M. Nikula, E. Saksman, C. Webb, Critical Mandelbrot cascades. Commun. Math. Phys. **325**(2), 685–711 (2014)
10. J. Berestycki, N. Berestycki, J. Schweinsberg, Beta-coalescents and continuous stable random trees. Ann. Probab. **35**(5), 1835–1887 (2007)
11. D.A. Dawson, Measure-valued Markov processes, in *École d'Été de Probabilités de Saint-Flour XXI—1991*, vol. 1541. Lecture Notes in Mathematics (Springer, Berlin, 1993), pp. 1–260
12. A. Dembo, Y. Peres, J. Rosen, O. Zeitouni, Thick points for planar Brownian motion and the Erdős-Taylor conjecture on random walk. Acta Math. **186**(2), 239–270 (2001)
13. A. Durand, Singularity sets of Lévy processes. Probab. Theory Relat. Fields **143**(3–4), 517–544 (2009)
14. A. Durand, S. Jaffard, Multifractal analysis of Lévy fields. Probab. Theory Relat. Fields **153**(1–2), 45–96 (2012)

© The Author(s) 2016
L. Mytnik, V. Wachtel, *Regularity and Irregularity of Superprocesses with (1 + β)-stable Branching Mechanism*, SpringerBriefs in Probability and Mathematical Statistics, DOI 10.1007/978-3-319-50085-0

15. N. El Karoui, S. Roelly, Propriétés de martingales, explosion et représentation de Lévy-Khintchine d'une classe de processus de branchement à valeurs mesures. Stoch. Process. Appl. **38**(2), 239–266 (1991)

16. A.M. Etheridge, *An Introduction to Superprocesses*, vol. 20. University Lecture Series (American Mathematical Society, Providence, 2000). ISBN 0-8218-2706-5

17. K. Fleischmann, Critical behavior of some measure-valued processes. Math. Nachr. **135**, 131–147 (1988)

18. K. Fleischmann, L. Mytnik, V. Wachtel, Optimal local Hölder index for density states of superprocesses with $(1+\beta)$-branching mechanism. Ann. Probab. **38**(3), 1180–1220 (2010)

19. K. Fleischmann, L. Mytnik, V. Wachtel, Hölder index at a given point for density states of super-α-stable motion of index $1+\beta$. J. Theoret. Probab. **24**(1), 66–92 (2011)

20. U. Frisch, G. Parisi, Fully developed turbulence and intermittency, in *Proceedings of the International Summer school Physics, Enrico Fermi* (North Holland, 1985), pp. 84–88

21. D.H. Fuk, S.V. Nagaev, Probabilistic inequalities for sums of independent random variables. Teor. Verojatnost. i Primenen. **16**, 660–675 (1971)

22. I.I. Gikhman, A.V. Skorokhod, *Introduction to the theory of random processes* (W. B. Saunders Co., Philadelphia/Pa.-London-Toronto 1969). Translated from the Russian by Scripta Technica, Inc.

23. R. Holley, E.C. Waymire, Multifractal dimensions and scaling exponents for strongly bounded random cascades. Ann. Appl. Probab. **2**(4), 819–845 (1992)

24. X. Hu, S.J. Taylor, Multifractal structure of a general subordinator. Stoch. Process. Appl. **88**(2), 245–258 (2000)

25. J. Jacod, *Calcul Stochastique et Problèmes de Martingales*, vol. 714. Lecture Notes in Mathematics (Springer, Berlin, 1979). ISBN 3-540-09253-6

26. J. Jacod, A. N. Shiryaev, *Limit Theorems for Stochastic Processes*. Grundlehren der Mathematischen Wissenschaften [Fundamental Principles of Mathematical Sciences], **288** (Springer, Berlin, 1987). ISBN 3-540-17882-1

27. S. Jaffard, Old friends revisited: the multifractal nature of some classical functions. J. Fourier Anal. Appl. **3**(1), 1–22 (1997)

28. S. Jaffard, The multifractal nature of Lévy processes. Probab. Theory Relat. Fields **114**(2), 207–227 (1999)

29. A. Klenke, P. Mörters, The multifractal spectrum of Brownian intersection local times. Ann. Probab. **33**(4), 1255–1301 (2005)

30. N. Konno, T. Shiga, Stochastic partial differential equations for some measure-valued diffusions. Probab. Theory Relat. Fields **79**, 201–225 (1988)

31. J.-F. Le Gall, L. Mytnik, Stochastic integral representation and regularity of the density for the exit measure of super-Brownian motion. Ann. Probab. **33**(1), 194–222 (2005)

32. J.-F. Le Gall, L. Mytnik, Stochastic integral representation and regularity of the density for the exit measure of super-Brownian motion. Ann. Probab. **33**(1), 194–222 (2005)

33. J.-F. Le Gall, E.A. Perkins, The Hausdorff measure of the support of two-dimensional super-Brownian motion. Ann. Probab. **23**(4), 1719–1747 (1995)
34. R. Le Guével, J. Lévy Véhel, Hausdorff, large deviation and Legendre multifractal spectra of Lévy multistable processes (2014). ArXiv:1412.0599v1
35. G.M. Molchan, Scaling exponents and multifractal dimensions for independent random cascades. Commun. Math. Phys. **179**(3), 681–702 (1996)
36. P. Mörters, N.-R. Shieh, On the multifractal spectrum of the branching measure on a Galton-Watson tree. J. Appl. Probab. **41**(4), 1223–1229 (2004)
37. L. Mytnik, E. Perkins, Regularity and irregularity of β-stable super-Brownian motion. Ann. Probab. **31**, 1413–1440 (2003)
38. L. Mytnik, V. Wachtel Multifractal analysis of superprocesses with stable branching in dimension one. Ann. Probab. **43**(5), 2763–2809 (2015)
39. E. Perkins, Dawson-Watanabe superprocesses and measure-valued diffusions, in *Lectures on Probability Theory and Statistics (Saint-Flour, 1999)*, vol. 1781. Lecture Notes in Mathematics (Springer, Berlin, 2002), pp. 125–324
40. E.A. Perkins, S.J. Taylor, The multifractal structure of super-Brownian motion. Ann. Inst. H. Poincaré Probab. Stat. **34**(1), 97–138 (1998)
41. M. Reimers, One-dimensional stochastic partial differential equations and the branching measure diffusion. Probab. Theory Relat. Fields **81**, 319–340 (1989)
42. R.H. Riedi, Multifractal processes, in *Theory and Applications of Long-Range Dependence* (Birkhäuser Boston, Boston, 2003), pp. 625–716
43. J. Rosen, Joint continuity of the intersection local times of Markov processes. Ann. Probab. **15**(2), 659–675 (1987)
44. S. Seuret, J.L. Véhel, The local Hölder function of a continuous function. Appl. Comput. Harmon. Anal. **13**(3), 263–276 (2002)
45. J. Walsh, An introduction to stochastic partial differential equations. Lect. Notes Math. **1180**, 265–439 (1986)
46. X. Yang, Multifractality of jump diffusion processes (2016). ArXiv:1502.03938v1

Index

© The Author(s) 2016
L. Mytnik, V. Wachtel, *Regularity and Irregularity of Superprocesses with*
(1 + β)-stable Branching Mechanism, SpringerBriefs in Probability and
Mathematical Statistics, DOI 10.1007/978-3-319-50085-0

Printed in the United States
By Bookmasters